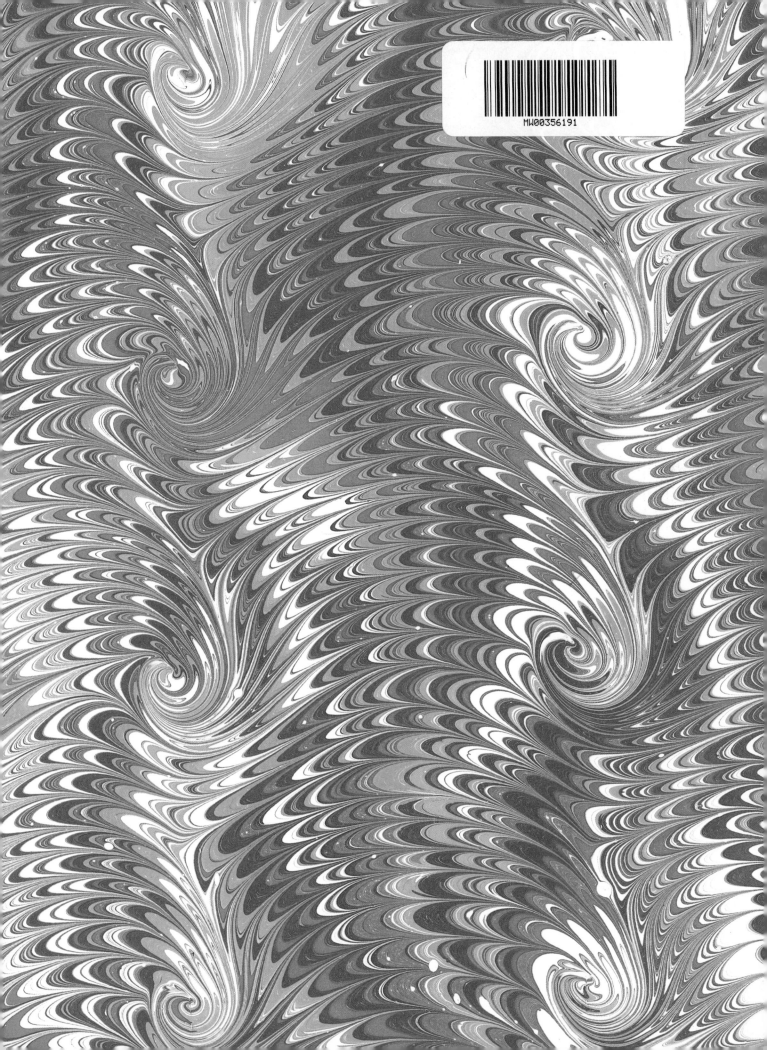

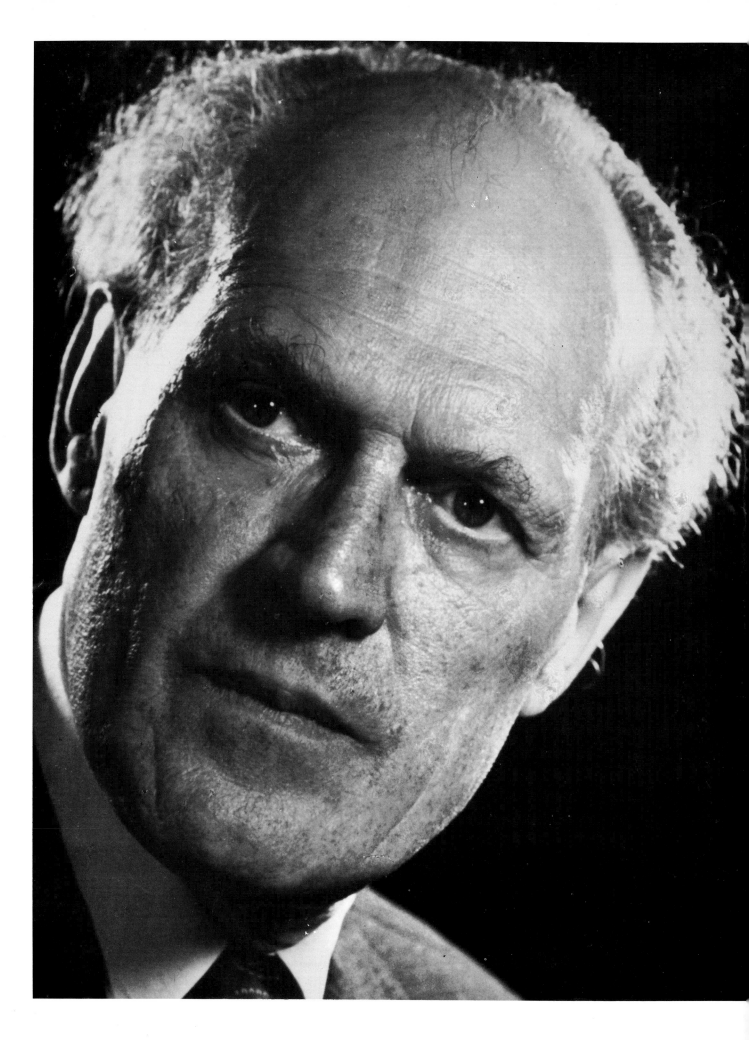

Willy Radinger & Walter Schick

SECRET MESSERSCHMITT PROJECTS

Schiffer Military History
Atglen, PA

Translated from the German by Ray Theriault.

Copyright © 1996 by Schiffer Publishing Ltd.
Library of Congress Catalog Number: 95-72486.

Printed in the United States of America.
ISBN: 0-88740-926-1

This book was originally published under the title,
Messerschmitt Geheimprojekte,
by Aviatic Verlag Peter Pletschacher.

We are interested in hearing from authors with book ideas on related topics.

Published by Schiffer Publishing Ltd.
77 Lower Valley Road
Atglen, PA 19310
Please write for a free catalog.
This book may be purchased from the publisher.
Please include $2.95 postage.
Try your bookstore first.

Contents:

Foreword 6

Introduction 7

Prelude 9

The Requirements 9
Early Work 16
Summary 19

A New Generation is Born 29

The P1101 Research and Fighter Aircraft 29
Interlude:
The P1106 as an Apparent Alternative 77
The Consequence: Project P1110, the
Classical Jet Fighter" of the Post-war Period 84
The Alternative: P1111/P1112 92
The Competition 102
Evaluation of the Designs 109
Summary 113

A Breakthrough, but not an End 116

Twilight of the Gods in Oberammergau 116
Help Yourself 118
Robert J. Woods continues the work:
From the P1101 to the Bell X-5 120

The HA-300 Light Fighter 134

A New Beginning in Spain 134
Finale in the Nile Valley 151

Appendix 165

Engine Variants 165
Technical Descriptions of the Jumo 004 and
HeS 011 Engines 167
Equipment/Armament 171
The Upper-Bavarian Research Institute
in Oberammergau 175
Tabular Summary of Development 177

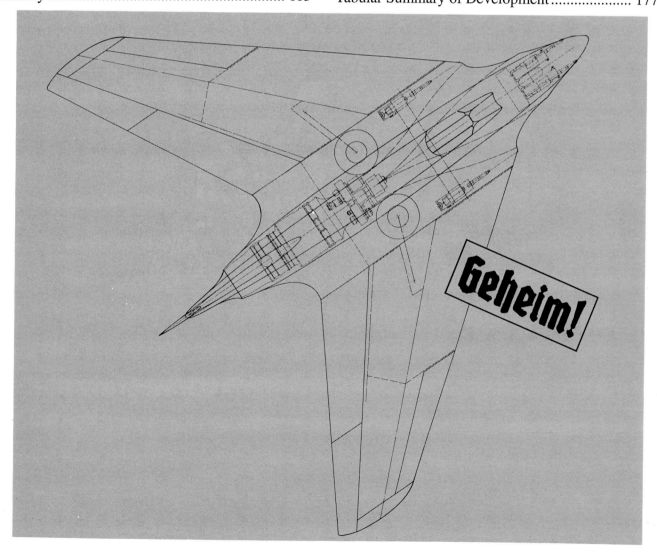

Foreword

The P1101 was Messerschmitt Construction's second jet aircraft after the Me 262. Designed as a fighter, the first test version of this type was to serve as an experimental aircraft to test swept wings and was expected to increase maximum air speeds considerably.

The airframe could have been completed in the spring of 1945, but the Heinkel engine planned for the aircraft was not yet available, and a Junkers' engine, which was originally intended to test flight characteristics, was more urgently needed for mass production. For these reasons, there was no opportunity to prove the aircraft's flight performance and characteristics.

After the war, the P1101 was brought to the USA and served there, among other things, as a testbed for similar developments.

Photos and data of this astonishing, yet understandably unknown aircraft with its futuristic technology next appeared in English-speaking literature. German readers only then learned of the existence of the Messerschmitt aircraft. This due to the post-war ban on aircraft production.

Since the P1101 was available only as a prototype, and had not even been flown up to this point; even the American press was limited to more or less short reports; quite in contrast to the numerous and extensive publications of other German aircraft prior to the end of the war. The same was true for the German press.

It is all the more welcome, therefore, that authors Radinger and Schick undertook the task of compiling documentation about this aircraft that was not even well known among experts, and thus present it to a wider circle of interested parties.

Following the war, details about German developments in the area of aircraft construction was a laborious undertaking; source material had to be assembled from English, French, and American archives, where all documents concerning German aircraft manufacturing had been stored in 1945.

The troublesome inquiries were worth the effort. There now exists comprehensive documentation where not only the developmental history with all the important data concerning the P1101 is gathered, but also an exceedingly ample photo collection, plus facts and figures about this aircraft. Of particular note are the numerous tabular comparisons with similar German and foreign aircraft from the war and post-war era.

While there are innumerable German technical publications about well-known, mass-produced Messerschmitt aircraft, the authors have been successful with this book in supplementing these works with the portrayal of the last, almost completed Messerschmitt aircraft.

Wolfgang Degel

Introduction

In many respects, the history of the Messerschmitt turbine-driven engine is, along with the works of the remaining German aviation industry and aeronautical research efforts, the developmental history of jet-propelled aircraft. Until far in the 1970s and still partly today, the jet-propelled aircraft has been or is in active service in air forces throughout the world, and its roots can be traced back to the Germany of 1944/45.

The Me 262 had been proposed by the end of 1938 and possessed a completely new jet engine and masterful aerodynamics. The successors to this aircraft received revolutionary aerodynamics which were a reflection of the higher speeds projected at the end of the war. The aerodynamics are, first of all, characterized by the swept-wing effect and all the aerodynamic solutions required to achieve these speeds such as normal swept-wings; crescent wings; delta wings; or swing wings, and later through such concepts as thin profile wings.

Messerschmitt's single-engine jet aircraft were not only known for their new type of engine, but also by the equally new aerodynamics which allowed Messerschmitt to develop aircraft with the highest performance and lowest weight. This also makes them fascinating to the technically interested, and this is why the central point of this documentation lies in the activities during 1944/45.

Jet-fighter projects P1073 and P1095, which do not actually fit within the outline above due to the tasking requirements (P1073) and aerodynamics (P1095), are included herein for the sake of completeness.

Except for the P1101 L in the appendix and the P1106 R, fighter aircraft with ramjets, pulse-jet engines, or rocket-powered engines are not covered in this book. The P1101 L study described should complete the history of the P1101, along with the remarks about the Bell X-2 and Bell X-5.

Furthermore, the single-engine jet developments by Alexander Lippisch from Messerschmitt, which suit the theme of this book, have also not been considered, since these were conducted by an independent development team which had hardly no influence on Messerschmitt and his research branch.

Had Professor Willy Messerschmitt fallen short in mass production in the area of the single-engine fighter aircraft, unlike the Bf 109 or the Me 262, he was revolutionary for the entire aerospace industry in the East and West through creation of the majority of the developments represented and the detailed solutions contained were substantially new.

Through these developments, engineers like Ludwig Bölkow during the war years or H. Langfelder and Gero Madelung in the Spanish period acquired considerable knowledge and experience, which they set aside for later in construction on one of the best combat aircraft in the world, the Panavia "Tornado."

Messerschmitt's profits for demonstrating the modern light fighter is a matter of concern for this documentation.

Moreover, from Messerschmitt's single-engine developments, unlike the Me 262, relatively few facts are known which are able to withstand scrutiny. The work contained herein should close that gap. And last but not least, the intentions of the authors, who both belong to a different generation of engineers, is to protect the splendid and technically scientific achievement of a group of engineers from being forgotten and, above all, to protect the achievements from falsification.

Willy Radinger, Walter Schick

The prelude

The requirements

The story begins in the early 1930s. One of the prerequisites was the realization of jet propulsion for aircraft. Turbine-, ramjet-or rocket-powered aircraft promised to immeasurably alter speed limits of propeller-driven aircraft, which were already foreseeable even at this time.

Such new engine types required an extensive new form of airframe in order to optimize present possibilities.

Much credit is due to the German engineers and scientists who immediately recognized this.

They created radically new aerodynamics with which it should also be possible, with equally radical propulsion, to carry out flights near the speed of sound and also supersonically.

The benefits and glory for the undertaking, which was partially accomplished under the most adverse circumstances, were reaped, above all, by Anglo-American and Soviet scientists after the war thus, the fate of the conquered.

A new engine: Since the beginning the 20th century, man has tried earnestly to build a useful gas turbine.

In 1921, Frenchman Guillaume's constructive and attainable proposal for a gas-turbine aircraft contained almost all the components of today's jet engine: multi-stage turbo compressor, combustion chamber, and multi-stage turbine.

In the early 1930s, the new form of propulsion was being developed almost simultaneously in both Germany and in Great Britain. In Germany it was mostly by Hans Pabst von Ohain, who realized his concepts while working with Ernst Heinkel in Rostock-Marienehe. In Great Britain, Air Force Officer Frank Whittle created the basis for revolutionary aircraft propulsion. Dr. Hans von Ohain and Frank Whittle embarked independently on the route to a centrifugal method of construction for the compressor and the turbine.

After innumerable preparatory works and attempts, which von Ohain conducted with his co-worker Max Hahn, they were successful in the spring of 1937 with the first experimental test-bed designated as HeS 2. One first result and a great success of the work was the world premiere of the jet aircraft on 27 August 1939. The engine, the HeS 3 of the small, high-wing He 178, was a highly developed HeS 2-engine with a thrust of 450 kilopounds (kp).

Heinkel and von Ohain remained true in principle to the centrifugal engine construction until the interpretation of the HeS 011. Engines with this type of construction could not prevail, however, against the axial designs from Junkers or BMW; a knowledge that, years later, would be painfully made by British engine design. Indeed, centrifugal engines clearly had their advantages, at least in the introductory phase of jet propulsion, which affected simple construction, uncomplicated production and, lastly, operational safety.

However, not only Heinkel was active. Since 1936 in Magdeburg, the leader of Junkers' Aircraft Development, Professor Herbert Wagner, was also busy with a jet engine with axial flow. Thanks to its small profile, this type of construction promised essential advantages at high flying speeds.

In 1939, Junkers-Motorenbau in Dessau received a Germany Air Ministry (RLM) contract for an en-

The first jet turbine engine aircraft in the world: the single-engine Heinkel He-178 V1.

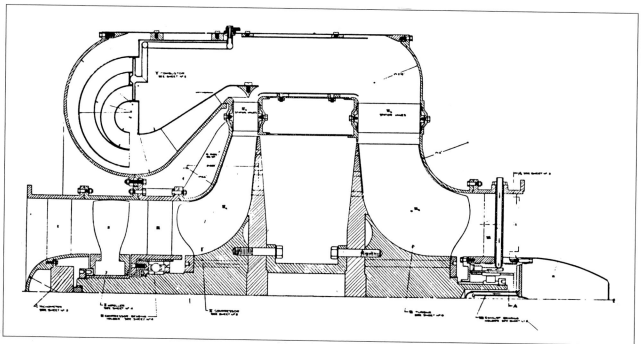

The He-178 V1 engine: Section of the Heinkel He S 3 engine.

gine capable of a 600kp full-throttle thrust. At the end of the year, a team headed by Dr. Anselm Franz took on the construction of the Jumo 004 engine. Where possible in engine layout, the engineers paid particular attention to using previously proven components. The compressor was a product of the Göttingen Aerodynamics Testing Laboratory (Aerodynamische Versuchsanstalt)-AVA from 1934/35, and Anselm Franz relied upon the practical experiences of the AEG-Berlin for the turbine blades.

When fully assembled, the resulting turbine jet-propulsion unit bore a striking resemblance to many successful jet engines of the postwar era. (See appendix)

The design of the Jumo 004 A was completed in the spring of 1940, and the first testing took place on October 11, 1940. The engine reached full RPM in December of the very same year.

Due to vibration fractures in the compressor intake blades, delays were experienced to such an extent that instruments were unable to measure the full 600 kp of thrust until the 6th of August. In December, the engine showed its hidden capabilities: after a successful 10-hour endurance run, the Junkers engineers were able to achieve a thrust of 1000kp.

In reaction to the achievements, the German Air Ministry (RLM) ordered 80 replicas of the Jumo 004 A for further engine development and for testing the airframe and aircraft operation. The engine made its first flight with a specially reconstructed Bf 110 test

aircraft and the Me 262 V3's first purely jet-driven flight with the Jumo 004 A engine took place on 18 July 1942 in Leipheim.

Because of an increasing scarcity of components had begun to have its effects, the Junkers engineers were forced to reconstruct a version of the engine that was relatively free of rationed materials. This mass-produced engine, designated the Jumo 004 B, began testing in the summer of 1943 and achieved a maximum take-off thrust of 910 kp.

Before the end of the war, the Junkers factory produced a total of 6010 units of this engine, which was the first engine to be built in mass-production.

Third in the "alliance" of German jet-engine pioneers was the BMW firm, which had taken up building a jet engines based on the gas turbine system. Just like Junkers, BMW in Berlin-Spandau also received a contract to develop an engine in the 600 kp thrust class. In contrast to the Junkers-Motorenbau contract in Dessau, BMW was obligated to adhere to measurements and weights as specified by the German Air Ministry (RLM) and were connected with the jet fighter development taking place at Messerschmitt.

Because of these circumstances, the BMW engine was built as close to mechanical limits as possible, and therefore was smaller in size and lighter in weight than Junkers' "competition."

At the beginning of 1941, the P3302 engine was placed on the test stand for the first time. At that time, however, the engine only reached a thrust of 150 kp. The engine's first flight in the Me 262 V1 on the 25th of March, 1942, also turned into a fiasco for the Berlin engine manufacturer.

It was clear to those responsible at BMW, that the thrust of 600 kp demanded of this engine was not to be attained. Dr. Hermann Oestrich and his team almost completely reconstructed the engine in the second half-year of 1942, and in this new construction, was easily able to exceed the required thrust with 800 kp in the middle of 1943. Once the German Air Ministry (RLM) awarded the contract, BMW could finally begin mass-production of what would now be called the BMW 003 jet engine. In August 1944, the first 100 mass-produced engines were delivered; altogether BMW built about 750 units, most of which were used, above all, in the Ar 234 and in the He 162 aircraft.

Building upon these efforts, Heinkel, Junkers, and BMW were able to develop a whole series of designs, some of which were given over to testing and served as inspirations for numerous further developments by German aircraft manufacturers.

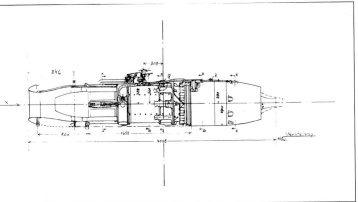

BMW 109-003 jet engine.

During the war years, the German engine industry suffered from a catastrophic lack of material. Those amounts of nickel, cobalt, and molybdenum required for heat resistant materials were simply not available. These were not required so much because their shortages hindered the actual construction, but more because of the first operationally ready axial engine would be able to demonstrate to the world its full potential.

Junkers engines were the chief influence for a series of Soviet designs. Dr. Oestrich and his team, which consisted to a large degree of his former BMW co-workers, created the extremely successful ATAR-series engines in France based on the BMW 003 turbine.

New aerodynamics: Leading aerodynamicists from Germany and Switzerland engaged in the legitimacy of supersonic flow in emulation of Austrians Mach and Salcher.

In addition to measuring increased air-flow witnessed by monitoring disruptions across various projectile-shape d forms as they crossed into supersonic speed, Prandtl, Busemann, Walchner, Ackeret and others also explored "supersonic flow on sharper airframes and wing surfaces."

Step by step, the laws and characteristics of supersonic flight became known. One of Prandtl's students, through Professor Th. von Karman, was able to bring these theories of supersonic flow to the United States in 1930. (It is therefore possible that the projectile-shape of the Bell X-1 had its foundations in an article which appeared in "Artilleristischen Monatshefte" (Artillery Monthly) in 1912; that the thin wing and control surfaces of history's first supersonic aircraft corresponds to one of the aerodynamic rules established by Prandtl.

In 1933, the two German aerodynamicists, Busemann and O. Walchner, published a scientific essay under the title "Profile Characteristics at Supersonic Speed."

By 1934, the famous space flight pioneer Professor Eugen Sänger was able to associate these successes in research with a rocket engine he had developed: In his essays "On the External Ballistics of Rocket Aircraft" and "Rocket Aircraft in Active Air Defense", where Sänger pointed the way to the supersonic interceptor.

The considerable advantage held by German aerodynamic researchers at this time over other nations became clear in the high-speed conference held in

held in Rome from 9 September to 6 October 1935:
Adolph Busemann from the Aviation Research Organization (LFA) in Braunschweig gave a lecture entitled "Aerodynamic Lift at Supersonic Speed." In his lecture, Busemann first publicly described the swept-wing effect to the aviation community.

Derived through mathematical calculations, the swept wing was considered an effective means of reducing resistance for supersonic flight for the first time.

In the year 1935, the time was obviously not quite ripe for such forward thinking, despite the existence of great aerodynmacists such as Th. von Karman, A. Grocco, G.I. Taylor and others. A few of the scientists, however, found the theory quite extraordinary and sketched drawings of these "Busemann Aircraft" on their banquet menus; aircraft outlines and wing shapes which, years later, would still be highly classified. For the time being, however, it appeared that Busemann's proposals would not receive the attention they deserved.

Hermann Göring founded the "German Academy for Aviation Research" (Deutsche Akademie für Luftfahrtforschung) on April 16, 1937, which, in contrast to the more application-oriented "Lilienthal Association", dedicated itself to theoretical questions concerning flight. At the third scientific session of this new organization, Messerschmitt conducted a much heeded lecture on the problems of high-speed flight and thereby called for a markedly increased research effort in the area of supersonic flight. As a board member of the Lilienthal Aviation Research Association (Lilienthal-Gesellschaft für Luftfahrtforschung - LGL), this successful aircraft manufacturer from southern Germany maintained almost constant contact with theoretical aerodynamics.

Almost a year later, A. Busemann once again urged the use of the swept-back, thin aircraft wing for subsonic and supersonic flight in the October issue of the magazine "Luftwissen." Since this scientific journal was most certainly read by interested individuals outside of Germany, it was particularly puzzling that the allies were surprised in 1945 when they came upon swept-wing models in wind tunnels.

Even the Lilienthal Association sponsored lectures at the end of the 1930s focused on problems of supersonic flight; on 14 December 1938, Dr. Max Kramer of the German Testing Institute for Aviation (DVL-Deutsche Versuchsanstalt für Luftfahrt) even reported

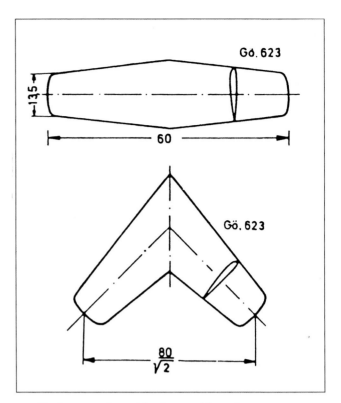

The first swept-wing model and a tapered-form wing for measurement in high speed wind tunnel of the Göttingen Aerodynamics Testing Laboratory (AVA).

on "The Influence of Mach Numbers and Boundary Layer Development on Profile Resistance." Through these events, it was quite clear to Messerschmitt that his fellow aviationists had become quite familiar with this new way of dealing with the problem.

But it was not until 1939 that Professor Albert Betz, Technical Director of the Göttingen Aerodynamics Testing Laboratory (AVA), took up his colleague Busemann's idea and began measuring resistance at subsonic speeds in conjunction with systematic research on the swept wing in the high subsonic range.

On September 9, 1939, the Reich Patent Bureau (Reichspatentamt) awarded secret patent Nr. 732/42 under the title "Aircraft with Speeds Approaching that of Sound" to the applicants, Adolph Busemann and Albert Betz.

After this more-or-less hesitant beginning, progress continued in quick succession. Toward the end of 1939, the AVA granted access to its research results to the aviation industry. Messerschmitt AG in Augsburg contracted the AVA for the systematic research of various swept wings and also had various fuselage and gondola shapes tested for high-speed wind resistance during the course of 1940. The measurements were carried out on an improved wing pro-

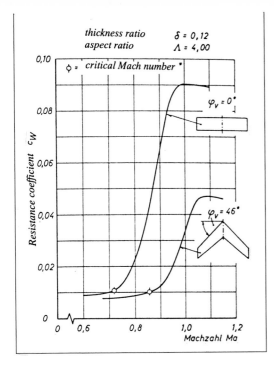

thickness ratio $\delta = 0{,}12$
aspect ratio $\Lambda = 4{,}00$

$\Diamond$ = critical Mach number *

$\varphi_v = 0°$

$\varphi_v = 46°$

Resistance coefficient c_W

Machzahl Ma

(Top diagram) Representation of the swept wing effect: Resistance coefficient for two wings with different angles of sweep in the transonic range.

rent test results — to include the works carried out by Messerschmitt.

One month later, the 1935 proposal made its way to the higher technical orders of the Messerschmitt firm: in technical report number 22/40, dated 21 October 1940, authors Bölkow and Schomerus summarized that "with an NACA 0012-64 airfoil and a wing swept from 33 to 38 degrees, it is possible to fly at speeds of at least 0.9M, or up to 1100 km/h at sea level without noticeable influence of compressibility to resistance.

However, clearly the cart was put well before the horse in this case, and a snake in the grass was buried in the details. Difficulties and disadvantages did not take long to surface. The main theme encompassed the following three problem areas:

• diminished slow-speed characteristics due to less total lift with all its negative side effects, such as increased take-off and landing speeds;

(Bottom diagram) Preliminary Swept-wing measuring program of the German Testing Laboratory for Aviation (DVL) in Berlin-Adlershof from December, 1940. Experimental model 4/6/9/12 and 13 are being evaluated for Messerschmitt AG.

file (instead of Dr. Busemann's Gö 623, Messerschmitt proposed a profile from the American NACA (National Advisory Committee for Aeronautics) classification, namely the NACA 0012-64) on sixteen different swept-wing models.

With the theme of "High Speed", a conference of the Lilienthal Association "Committee for General Flow Research" took place on the 3rd of September in Braunschweig and on the 26th and 27th of September 1940 in Göttingen, Representatives of the research and aviation industry were informed about cur-

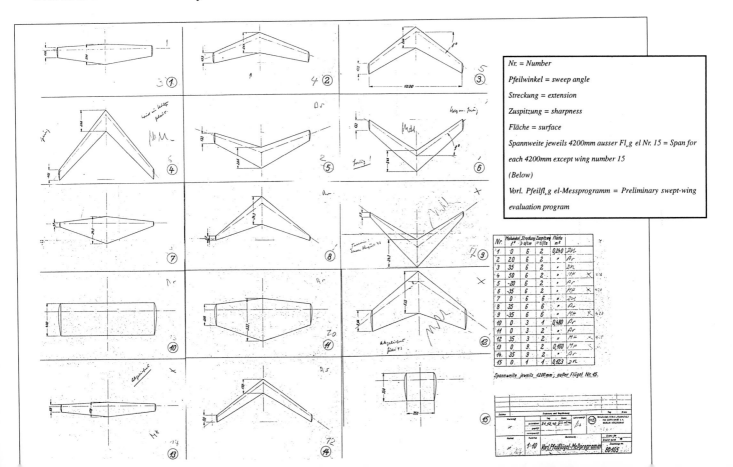

Nr. = Number

Pfeilwinkel = sweep angle

Streckung = extension

Zuspitzung = sharpness

Fläche = surface

Spannweite jeweils 4200mm ausser Fl,g el Nr. 15 = Span for each 4200mm except wing number 15

(Below)

Vorl. Pfeilfl,g el-Messprogramm = Preliminary swept-wing evaluation program

Inner and outer leading-edge slats of the Me-262 (extended).

• area of flow separation on the outer wing as a result of boundary layer displacement;

• constructive disadvantages, such as the emerging high rear center of pressure position, in comparison to the straight wing with essentially larger torsion load, which can lead to a reversal of the aileron effect.

The engineers and scientists involved immediately accepted the task of eliminating of or at least partial removal of these defects by using the means which technology offered at that time. If they could not find a way through already existing technology, they created it. The courage and creativity with which they entered this uncultivated technical territory is impressive. A few examples will help stress this;

• In the Bf 109 and Me 262, Messerschmitt used almost the entire wing span for improvement of the slow-flying capacity leading-edge flaps (slats). Through creation of the new generation of jet fighters, Messerschmitt naturally retained this feature which, after the war, could be seen in many high performance aircraft, especially those from the aircraft manufacturer North-American;

• a different, essentially more radical proposal, likewise for improving low-speed flight characteristics, came from the section "L" of the Messerschmitt Firm. Dr. Alexander Lippisch, who was himself active since 1939 with his own section in Augsburg, also named "L", created a suitable delta wing as a special form of swept wing for not only low and high speed

flight, but he applied for a patent in 1941 with the title "Swept Wing With Rotating Wing Tips." Therein, Lippisch was already describing the variable geometric wing, which displayed two bearing supports outside the fuselage; an arrangement which was used after the war by, above all, Soviet designers like Sukhoi and Tupolev.

By the end of the war, Messerschmitt and the Hamburg Research Department of Blohm & Voss worked on studies for tilt-wing aircraft. Indeed, both firms went in different directions from Dr. Lippisch: Blohm & Voss' P.202 project possessed an solid wing, with only a pivot bearing being provided, and resulting in a forward-swept and reverse-swept wing. At the beginning of the eighties, NASA tested this unusual idea with the Ames AD-1, and as far as is known, the attempts at flight proceeded very successfully.

Likewise, the Messerschmitt-solution stored a divided, tiltable wing in the fuselage: both wing halves of the wing assembly unit would pivot into what today is the usual swept-back manner.

In low-speed flight the aircraft with swept-back wings has problems not only with buoyancy, but it also has the dangerous tendency of nose-diving to one side. But even at a low angle-of-wing setting, local flow separation occurs at the wing tips which, due of the non-symmetrical appearance, would inevitably lead the aircraft to the aforementioned pitching. The cause of this is the movement of the boundary layer over the wing and across in the direction of flight to the wing tips. To prevent this, Messerschmitt's aerodynamicists suggested the following:

• Aside from using slats in the outer wing area (see P1111), they also explored a wing with different sweep settings (crescent wing) and anticipated using this for the very first P1101-studies and for a bomber project;

• One feature they created was a wing with a proportional thickness of 8 percent less on the inner wing than that of the outer wing (12 percent). Aside from structural advantages (torsional rigidity!), the aerodynamic decalage was supposed to maintain lateral control in high speed flight (prevention of the notorious "flutter"), and boundary layer movement in low-speed flight, thereby avoiding the aforementioned flow separation.

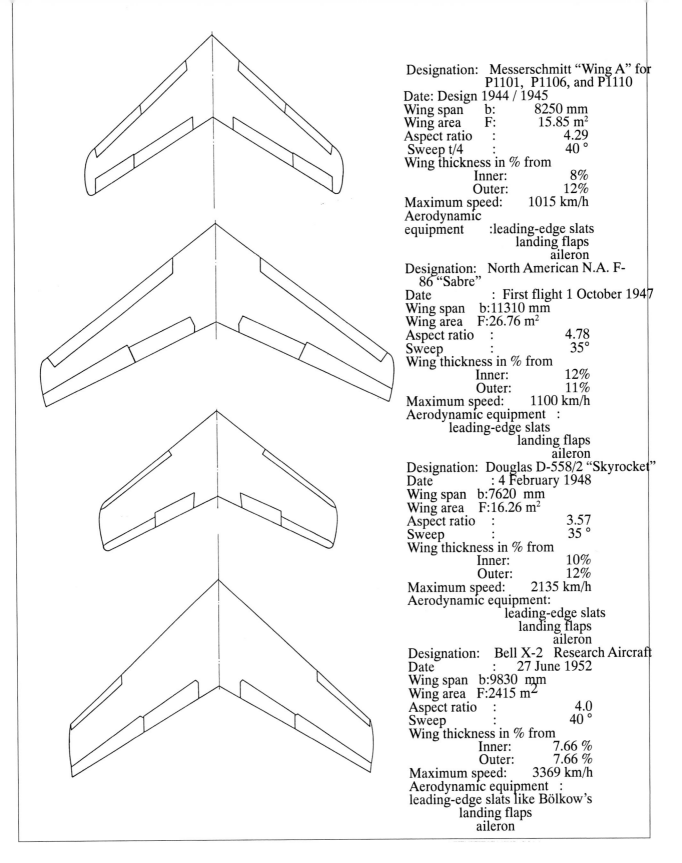

Designation: Messerschmitt "Wing A" for
                    P1101, P1106, and P1110
Date: Design 1944 / 1945
Wing span   b: 8250 mm
Wing area   F: 15.85 m²
Aspect ratio   :   4.29
Sweep t/4   :   40 °
Wing thickness in % from
                 Inner: 8%
                 Outer: 12%
Maximum speed: 1015 km/h
Aerodynamic
equipment   :leading-edge slats
                    landing flaps
                    aileron
Designation: North American N.A. F-
      86 "Sabre"
Date   : First flight 1 October 1947
Wing span   b:11310 mm
Wing area   F:26.76 m²
Aspect ratio   :   4.78
Sweep   :   35°
Wing thickness in % from
                 Inner: 12%
                 Outer: 11%
Maximum speed: 1100 km/h
Aerodynamic equipment   :
          leading-edge slats
                    landing flaps
                          aileron
Designation: Douglas D-558/2 "Skyrocket"
Date   : 4 February 1948
Wing span   b:7620 mm
Wing area   F:16.26 m²
Aspect ratio   :   3.57
Sweep   :   35 °
Wing thickness in % from
                 Inner: 10%
                 Outer: 12%
Maximum speed: 2135 km/h
Aerodynamic equipment:
                    leading-edge slats
                          landing flaps
                          aileron
Designation: Bell X-2   Research Aircraft
Date   : 27 June 1952
Wing span   b:9830 mm
Wing area   F:2415 m²
Aspect ratio   :   4.0
Sweep   :   40 °
Wing thickness in % from
                 Inner: 7.66 %
                 Outer: 7.66 %
Maximum speed: 3369 km/h
Aerodynamic equipment :
leading-edge slats like Bölkow's
                    landing flaps
                          aileron

A wing for many uses: Messerschmitt wing "A" from 1944/45.

A wing constructed as such, with a slat over almost the entire wing span, was to be tested with the P1101 experimental aircraft. From the outset the wing designated concept "A" was to be used for the P1101 in mass production and for the P1106 and P1110 designs.

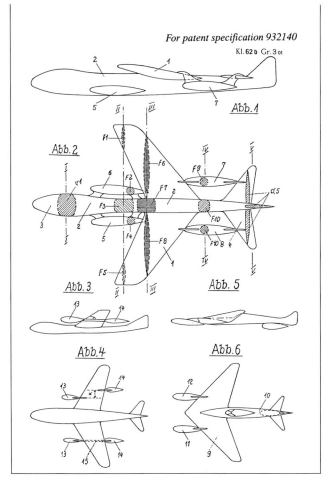

For patent specification 932140

Abb. = Illustration

The type of construction with aerodynamic and/or geometric decalage ("isobaric concept") is nowadays preferred by many aerodynamicists through the realization of swept-wing construction. It has the advantage of having the wing kept clear of other aerodynamic equipment (such as boundary layer fence, saw tooth, wing slot, turbulence stimulator, etc).

The Messerschmitt Research Institute owns these aerodynamic concepts and other aerodynamic solutions as well (for example: new, thin aircraft profiles with leading-edge flaps in lieu slots) with the most valuable and scientific spoils of war going to the USA.

The aerodynamicists of the Firm Arado had hoped for similar effects of the boundary-layer effect. They had explored the crescent wing as well as the swept wing, which displayed a larger depth on the wing tip than on the wing root. After the war, American engineers implemented the XF-91, which flew faster than the speed of sound in December 1952, however, they did not further pursue the idea.

Dr. Wolfgang Liebe of the German Testing Institute for Aviation (DVL) came up with a further solution to the problem: he wanted to prevent boundary-layer movement by fastening web plates on the wings in the direction of flight; the boundary-layer fence was already invented.

Besides the swept-wing effect there was another grand aerodynamic discovery during the 1940s: the law known as the "rules for wing surfaces" of aerohydrodynamics, whereupon in the area of transsonic and supersonic flight, the resistance of an aircraft is dependent on its cross-sectional distribution.

The first attempts at this realization are surely found in the analysis already covered on supersonic-speed artillery shells at the beginning of the century.

During the war, some German aircraft manufacturers (above all Junkers, Heinkel, and Messerschmitt), with the aid of the Research Institute, explored the optimal assignment of aircraft components to one another in order to minimize resistance in "high-speed flight." As a result of this analysis, in 1944 Junker's cohorts Hertel, Frenzel and Hempel were able to grasp the laws discovered in the handy "rule" and received Patent Specification Number 932410 for their work. From the beginning, Messerschmitt's Project Bureau attempted to heed these obviously expanding laws through interpretation of single-engine fighter aircraft design. The later designs of the P1110 and, above all, the P1112 show a knowledge of the rules for wing surfaces by their shape.

First works

The world's first jet, the Heinkel He 178, was a single-engine experimental aircraft. The obvious advantages and the thoroughly sufficient thrust performance (600 kp) of the jet-power plant apparently permitted an almost seamless transition from the single-engine aircraft to the single-engine jet fighter.

At the end of 1938/beginning of 1939, the very first designs of the Messerschmitt Project Bureau in Augsburg for Project 1065 emerged, from which the Me 262 would later materialize: the Messerschmitt Project Bureau therefore only displayed one turbine jet-propulsion unit.

Indeed, the project engineers tried to steer clear of the long air ducts of the He 178. This is obvious by the proposed double fuselage aircraft of the later D.H 100 "Vampire" and on another aircraft with a lower hanging engine similar to the later 1095 project. However, the technical data for the BMW engine P3302, which was to be included, quickly changed to their disadvantage.

For this, Engineer Degel, then heavily involved in the development, stated: "With the Me 262, we have begun single-engine design and we've explored all conceivable accommodations for the engine. This was unexpectedly the first initial stage for the P1101. Only after the specified engine became heavier, thicker and used less thrust than was expected were we compelled to provide two engines."

After BMW developed the new, up-and-coming P3304 engine in the summer of 1940, Messerschmitt once again risked designing a very small, single-engine fighter which, due to its specifications and the resultant explanation, takes a special place in this document.

The P1073 B Fighter: P1073 is the general project designation for the specification which was much discussed during the winter of 1939/40 and was clarified in series of preliminary projects.

The goal of the strategists was to engage long-range aircraft in a "battle over the Atlantic." In an early phase the underlying thought of the investigations was to observe the convoys as they gathered and to clutter them by dropping single bombs. That progressed to expand the aircraft tasking to "isolated bombing of the U.S. east coast", whereby one expected a strong contingent of opposing forces and looked for a means to repulse this attack.

In the 1930s, after Soviet designers already developed the idea of the "flying aircraft carrier" (Project "Sweno"), Messerschmitt's future project considerations went in a similar direction. Thus, in August 1940, a study in long range fighter aircraft emerged. The aircraft was itself supposed to guide its own escort in the form of a manned fighter.

The project as a whole received the designation "P1073-Long-Range Bomber-Aircraft Carrier with Three Sea Launched Fighters", whereby the gigantic, 63 meter launching aircraft was designated P1073, and the small parasite fighter attached, the P1073 B.

Mathematically the aircraft was an eight Jumo-223 motor-driven, long-distant bomber with the three parasite fighters attached, with a 6000kg bomb load, a fuel range of 16 000 km and a take-off weight of about 128 tons.

It was planned that the three escort aircraft would refuel with a pilot in the fuselage of the carrier aircraft while on a guide beam from a hatch. In operational use, these three aircraft would be extended and dropped. Even with safety being considered, the planned retraction maneuver, accomplished with a type of tow rope, would be a weak point of the aircraft's concept.

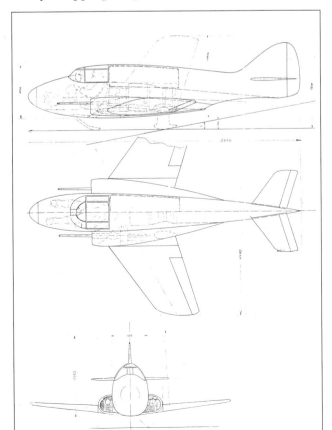

Left photo: Original design of the P1073 B parasite fighter from 13 August 1940 (left).

Right photo: The American towable fighter McDonnell XF-85 "Goblin" (above).

While no design from the P1073 A launching aircraft survived, some specifications and the design from 13 August 1940 gave a relatively clear picture of the anticipated parasite fighter. The low-wing monoplane with the high, oval fuselage section possesses a faired cockpit canopy and a rudder which clearly reminds one of the Me 209.

The wings are foldable due to the better storability, therefore when loading it onto the launching aircraft, only a transport cross section of 1.73 meters x 1.71 meters is required.

A wing without lift-increasing devices and with a laminar profile has a wing sweep setting of 35 degrees at the t/4-line. The ailerons extend over the trailing edge of the wing by approximately 80% and take up about 1/5 of the chord length.

In lieu of landing gear there are two rotatable nose skids and a pair of extendible landing skids for weapons cover discharge.

Together with the Lippisch works from the summer of 1940 (P01 -112/113/114), the 1073/B Project represented the first serious proposal for an aircraft with which conscious use of Busemann's swept-wing effect formed a basis.

After Messerschmitt failed to follow through with the idea, the Americans once more continued with development and construction of the McDonnell XF-85 "Goblin." From the earthy looking, oval-shaped aircraft, one could find three of them in the bomb bays of the gigantic Convair B-36.

However, before the first flight of the small fighter took place on August 23, 1948, the U.S. Air Force had changed its strategy; as a consequence of this, the thirty aircraft that were ordered were scrapped.

Nevertheless, they carried out flight tests of both prototypes, 46-523 and 46-524, in Muroc with the Boeing EB-29 as carrier aircraft right up until the 18th of March, 1949.

Results of the flights were everything but encouraging. After the outstanding test pilot, Edwin Schoch, even had difficulties with the strange bird, the question was raised: how well would the average Air Force pilot manage?

The inferior performance of the XF-85 compared to both traditional interceptors and the more perfected surface-to-air missile systems underlined the correctness of the preceding Air Force decision based on the test flights.

	Messerschmitt P1073/B	McDonnell XF-85 "Goblin"
Primary function	single-seat parasite fighter	single-seat parasite fighter
Powerplant	1 x BMW 3304 (BMW 002) with 600 kp static thrust	1 x Westinghouse J-34 -WE7 with 1361 kp static thrust
Length	5900 mm	4534 mm
Height	1810 mm	
Transport-device	1.71 m x 1.73 m	3.25 m x 1.645m
Wing span	4400 mm 6440 mm	
Wing area	6.5 m^2	8.36 m2
Aspect ratio	approx. 3.0	approx. 5.0
Wing sweep	35 degrees/t/4-line	37 degrees leading edge
Empty weight		1697 kg
Flight weight	1620 kg	2064 kg
Wing load	approx. 250 kg/m2	247 kg/m2
Landing gear	skis	skis
Max. speed	935 km/h theoretically	attained: 583 km/h 1068 km/h theoretically
Service ceiling		15,240 m
Fuel capacity	250 liters	424 liters
Max. flight time	31 minutes (with 50 % thrust)	30 minutes
Armament	2 x MG 151/20 (20 mm)	4 x 0.5" MG (12.7 mm)

A new approach

Conventional fighter and high-altitude fighter P1092: Between the years 1940 and 1943, when the Messerschmitt Manufacturing Plant developed a whole series of novel aircraft and manufactured some of them, work on the Me 262 outstanding fighter aircraft, and designs for a single-engine fighter ended up on the back burner.

In the meantime, both the military and economic situations of the time required new considerations to come to the forefront. And the development stage attained with the turbo-jet engine, especially during rationing and material shortages, was instrumental in the realization of the single-engine aircraft as a tangible alternative to the twin-engine concept (Me 262, He 280, Ar 234).

In the spring of 1943, first off, Focke-Wulf and Messerschmitt, after inquiries and encouragement from the German Air Ministry (RLM), began work with single-engine designs.

For Messerschmitt, who at that time was the only aircraft manufacturer in the world, and also who, aside from numerous aerodynamic innovations, also transformed three revolutionary new engines (turbine — jet, rocket, and push pod engines) into mass production, he offered an elaborate single-engine design over and above the possibility of a technical, performance rated and economical aircraft comparable to the Me 262. Still, this "competition" was as much a result of Technical Report Number 90/43 "Performance Comparison of Turbo-Jet Fighters" in favor of the twin-engine fighter.

As often is the case, Project P1092 was a collective name for a whole group of studies and deliberations for aircraft with various roles.

Technical data:	
Designation	P1092 design XVIII/44 (1) from 8 June, 1943
Primary function	single-engine fighter
Crew	1 pilot
Powerplant	1 x Jumo 004C with 1015 kp thrust and 1200 kp with afterburner
Wingspan	8500 mm with inner wing highly swept and outer wing slightly swept
Wing area	13.5 m2
Aspect ratio	l = 5.36
Dihedral angle	= 10 degrees
Fuel	900 kg or 1200 liters when g = 0.75
Take-off weight	3850 kg
Wing load	285 kg/m2 (take-off)
Military armament	2 x Mk 103 2 x MG 151/15
Maximum speed	910 km/h at altitude of 6 km
Rate of climb	18.3 m/second (take-off)
Maximum range with climb and cruising distance	1025 km
Maximum flight duration	2 hours at altitude of 9 km
Service ceiling	11,600 m
Take-off distance	700 m

From these mostly estimated figures it's already seen that the later 1101 Project got its starting point with this development work.

For the moment, the first designs encompassed twin turbo-jet aircraft with the designations P1092 A/B/C/D and E, with a rocket engine or with two Argus tail pipes. The obvious military use was also abundantly clear: fighters, high-altitude fighters, interceptors, night fighters, destroyers, high-speed bombers, Stuka and torpedo aircraft.

These studies made sense with the basic design derived from the Me 262, thereby achieving, on one hand, a reduction in expenditure, and on the other hand the utmost operational diversity and a certain liberty in using the engine.

It was only with the "P1092 fighter with 1 Jumo-109-004C turbo-jet engine" from the 8th of June, 1943, that the initial passage to the single-engine fighter was clearly pursued.

The aircraft represented by Eberhart, Hornung, and Voigt, designated XVIII/44 (1), largely met the specifications of the explicitly built P1092 from the 3rd of July, 1943, except for the slight wing sweep.

After this start, several suggestions followed in the next few days in which the basic design from the 8th of June was altered. Thus, on June 11, 1943, with an unregistered design, the "single-engine aircraft with Jumo 109-004C" engine received a smaller wing with a continuous leading edge, and on a likewise unregistered combination from June 26, 1943, the P1092 exhibited the wings and stabilizer of the Me 262; furthermore, an unfinished design from June 30, 1943 shows a somewhat enlarged P1092 with the following weaponry: 2 x Mk 103, 2 x Mk 108 and 2 x 250 kg bombs.

At the start of July, it became clear that the aircraft proposed would be used for the aforementioned experiments comparing the tailless P20, derived from the Me 163, and the twin-engine Me 262 and even the P1092.

Hornung and Voigt summed up their results in Technical Report Number 90/43 as follows:

Summary:
The fighter with two turbo-jet engines (Me 262) was compared to the fighter with one turbo-jet engine; the latter will be tested under normal construction (P1092) and in tailless design. The aircraft is compared using equal operational uses (equipment and weaponry). The landing speed for both single-engine fighters is the same, which is 8km/h higher than that of the Me 262.

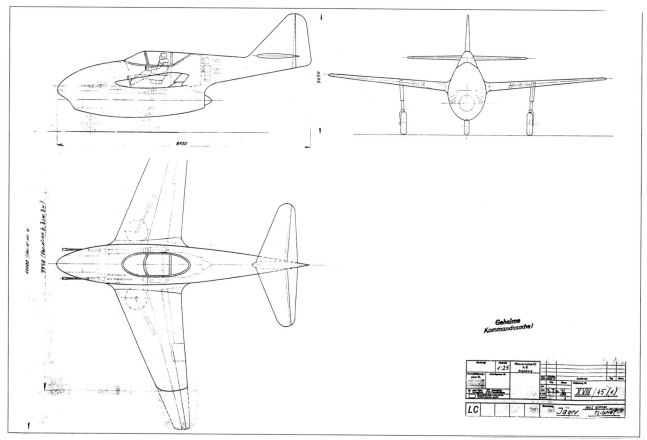

P1092 a /b/b H "single-engine fighter in normal construction" with designation XVIII/45 (1) from 3 July, 1943.

With the single-engine conventional aircraft, a modification of the weight and aspect ratio is accomplished. Weight alleviation is possible with the same success due to the tailless construction, and aspect ratio modification was not explored with the latter since, in hindsight, because of the short deadlines under all these circumstances, the aerodynamic data of the Me 163B had to survive.

Result:

I) P1092 a / P20

Comparison of the single-engine fighters with equal aspect ratio shows the following:

The traditional design is considered superior by an average of approximately 25 km/h in maximum speed.

The tailless design is superior in the following performance areas:

Climb rating in higher altitudes: maximum altitude difference - 450 meters.

Duration of flight: no difference at ground level, 15 minutes = 22% at altitude of 11 km.

Range: no difference at ground level up to altitude of 9 km, 55 km = 7% at altitude of 11 km.

Turning radius: tightest curve without altitude loss at an altitude of 6 km, r = 780m against 1000m.

Both designs are equal in start and landing distances as well as descending speed.

II) P1092 b/P20

Comparison as above, however with increase of wing span of the conventional aircraft (large wing tip).

The conventional aircraft is considered superior in all performance areas. The superiority totals:

Maximum speed   10 km/h
Maximum altitude   400 m
Range, no difference at ground level, up to altitude of 11 km, 30 km = approximately 3.5 %
Duration of flight, no difference at ground level, at altitude of 11 km 6 minutes = 6.5 %
Take-off distance rd. 50 m = approximately 10 %
Turning radius (as calculated above) 706 m compared to 780 m
Descending speed 7.6 m/second compared to 9.9 m/second
Landing speed 165 km/h against 176 km/h.

III.) P1092 b/Me 262

The comparison between the best single engine

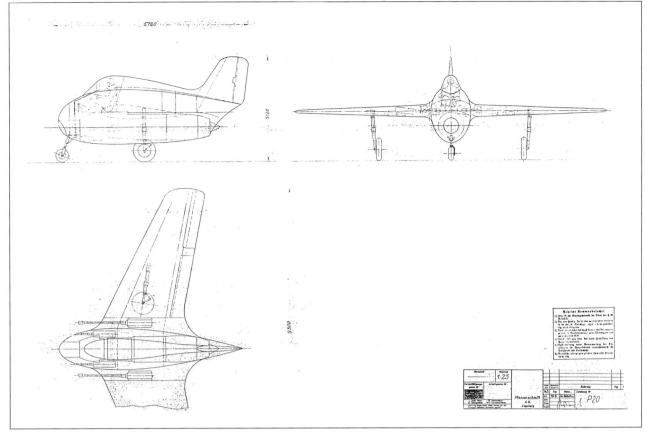

The P20 single-engine fighter in its "tailless design"; Original design by Dr. Wurster.

fighters (increased wing span) and twin-engine (Me 262) resulted in the following:

The twin-engine aircraft is, of course, superior to the single-engine aircraft in altitudes up to 6 km and by speeds of approximately 20 km. However, in altitudes over 6 km it is considerably superior. Otherwise, the twin-engine aircraft is weaker in the following areas:

Take-off distance: . 690 m compared to 635 m
Turning radius: ...... 1000 m compared to 700 m
Descending speed: 9.0 m/second compared to 7.6 m/second
Landing speed: ...... 184 km/h compared to 165 km/h

In all other performance areas, the twin-engine aircraft is decidedly better in the following areas:
Maximum speed at altitudes over 6 km
Rate of climb at ground level 3.7 m/second, and in maximum altitude approximately 400 m
Range:
at ground level approximately 120 km = 30%
at altitude of 11 km approximately 380 km = 40%
Duration of flight:

at ground level for 18 minutes = 30%
at altitude of 11 km for 42 minutes = 40 %

Moreover, in regards to overload capacity by increasing fuel capacity or taking on bombs, the twin-engine aircraft is absolutely superior to the single-engine aircraft.

The single-engine aircraft naturally requires only half the number of engines and lesser amounts of fuel, and requires 55% less material expenditures and is approximately 85-90% the production cost of the twin-engine aircraft.

In purely mathematical terms, the contrast reads: (see table on page 22).

Besides the information for the Lippisch P20 and the Me 262 , the table has a very comprehensive overview of technical data for the P1092 a standard design, for the P1092 design with enlarged wings, and of the P1092 b H high-altitude fighter version.

Between Messerschmitt and the RLM in the summer of 1943, there was still no concrete intent to build any of the P1092 aircraft; but the experiences with this project proved very worthwhile when the development contract from the RLM finally arrived for planning and construction of the P1101.

The results of the comparison are worth taking notice due to the fact that, at this time, the studies made no valid pretenses. This was simply a comparison between three

different concepts from the Messerschmitt camp, and the picture had already changed approximately one and a half years later following the inclusion of a more powerful engine and applying the latest results from the aerodynamic research in favor of the single-engine design, at least for those aircraft designated as "fighter planes."

In connection with Technical Report Number 90/43, Hans Hornung attempted to find a better design

	P1092/A	P1092/b	P1092/bH	P20	Me262
		Dimensions			
Wing surface F(m²)	12.7	14.45		17.3	21.6
Wing span b(m)	7.75	10.00		9.3	12.4
Aspect ratio	4.75	7.00		5.00	7.05
Average profile	0011.3	0011.3		13.4%	0011.3
Wing surface 0n(m²)	20.6	24.2		31.0	32.1
Fuselage length	8.31	8.31		5.75	10.60
Fuselage surface 0pm(m²)	24.3	24.3		18.7	31.1
Tail surface 0LW(m²)	9.4	9.4		2.8	13.7
Gondola surface 0 (m)²	-	-		-	18.4
Total surface area	54.3	57.9		52.5	95.3
		Weights			
Airframe (kg)	919	985		917	1655
Engine	874	874		864	1792
Empty weight	234	234		234	236
Armament	434 2MK103 + 2MG151/15	434 2MK103+ 2MG11/15	1MK103 250 or 2MG151/15	434	414
Armor plating	165		165	140	230
Total weight	2626	269	2343	2589	4327
Crew	100	100	100	100	100
Fuel	750	750	750	750	1831
Munitions	188	188		188	264
Take-off	3664	3730	3193	3627	6522
Expenditure for take-off and engine warm-up	75	75	75		
Lifting weight	3589	3665	3118	3552	6372
Landing weight	3251	3317	2780	3214	5531

		P1092/a	P1092/b	P1092/bH	P20	Me262
Take off	Wing surface F(m²)	12.7	14.45	14.45	17.3	21.7
	Wing loading for lift	28.7	253	216	205	294
	With 1200 kg take-off thrust	0.334	0.328	0.385	0.338	0.376
	Take-off run when = 0.07 take-off run (m)	690	635	440	685	690
In flight	rate of climb					
	at ground level	18.4	18.3	22.4	18.5	22.0
	at 6 km altitude	9.6	10.3	12.9	10.0	12.5
	at 10 km altitude	2.7	4.0	5.8	3.4	5.2
	Climb time to 6 km					
	T (min) 7.3 to 10 km	7.2	5.7	7.2	6.0	
	(minute) = 540 km./h	18.8	16.9	13.1	17.9	14.0
	Service ceiling (km)	11.2	12.1	12.9	11.6	12.5
	High-speed flight with 100% thrust					
	at ground level 872	858	852	846	850	
	at altitude of 6 km	931	914	926	905	933
	at altitude of 10 km	890	892	917	866	935
	At altitude equal to 1 meter/second climb rate	832/11	795/1.6	810/12.6	785/11.3	876.12
	Range with cruising and climbing					
	at ground level	390	394	420	385	515
	at altitude of 6 km	665	695	739	669	930
	at altitude of 10 km	855	925	1040	885	1290
	at 1 meter/second climb rate	870/11 km	970/11.8	1210/12.6	940/11.3	1430/12
	Duration of flight with climbing and cruising time of an hour					
	at ground level	0.21	0.81	0.84	0.81	1.11
	at altitude of 6 km	1.15	1.35	1.151	1.29	1.87
	at altitude of 10 km	1.26	1.59	1.85	1.50	2.23
Landing						
	Best drag-to-lift ratio with maximum landing load	12.8 / 264	15.48 / 230	15.48 / 193	14.00 / 186	12.65 / 255
	Descending speed at approx 0.9 maximum in 0 m without flaps meter/second	6.85	4.67	4.28	5.30	6.23
	Descending speed at 0 m with extended landing flaps Wg (m/sec)	9.90	7.61	6.95	9.92	9.00
	Maximum landing coefficient Speed	1.75 / 176	1.75 / 165	1.75 / 151	1.25 / 176	1.57 / 184
Turning	Tightest bank	3320	3320	2780	3215	5550
	At ground level tightest turning radius r (m)	398	316	288	335	402
	shortest turning time for full circle (in seconds)	28	21	17.7	23	24.2
	at altitude of 6 km tightest turning radius r(m)	1000	706	637	780	1010
	shortest turning time for full circle (in seconds)	78	58	45.5	61	63.2

for the single-engine fighters. Similar to the later P1101, he had the cockpit moved to the rear. On the aircraft designated XVIII/46 from the 16th of July, 1943, in the aircraft nose next to the intakes, the four Mk108's and behind them in the center-of-gravity range (envelope), the Jumo-004 engine are housed with the entire fuel supply (1250 liters). The cockpit lies far behind and over the engine exhaust directly before the stabilizer.

Three days later, on July 19, 1943, a further over-view showed a totally opposite cockpit: on the designation XVIII/47, the pilot is completely forward and positioned over the air intakes.

A further proposal from the P1092-series, designated Number XVIII/48 from July 20, 1943, showed an aircraft that matched the Me 262 from proposal XVIII/45 (1) from the 3rd of July, 1943, right down to the wings.

With this, project series number 1092 ends. The path to the P1101 had now literally been etched.

In November 1943, in Chalais-Meudon near Paris, wind tunnel testing with a wing span of 10.0 meters and a sweep of 35 degrees was planned. The results of these tests were recorded in two reports:

Report Number 7822 "Lift Measurements with Wings Swept Rearward at a 35 Degree Angle in a Wind Tunnel in Chalais-Meudon" and the second, report Number 7852 "Further Results from Measurements of a Wing Swept to a 35 Degree Angle in the Wind

Tunnel at Chalais-Meudon." One can the proceed on the assumption that these tests and their results from French specialists working in Chalais-Meudon did not remain unseen.

On June 23, 1948, in France, or to be more exact, in Chatillon/Bagneux, an aircraft arose for the first time whose measurements and weights matched exactly those of design project "1092 a" from July 3, 1943. The French airplane used a wing sweep of 35 degrees (P1092: 20 degrees) and for propulsion, a Junkers Jumo 004 B-2 engine.

This machine dealt with swept-wing research aircraft VG-70 of the Arsenal de'l Aeronautique. Jean Galtier was responsible for construction of this design, and he also later brought forth the Nord "Griffon II."

View of the French swept-wing experimental aircraft Arsenal VG-70.

It can be safely said that for the construction of this aircraft, German documents were readily available. Mentioning this aircraft type should serve to illustrate the outstanding work of the Messerschmitt Project Bureau. This example shows how near Messerschmitt came in calculating the lower boundary layer to the performance actually achieved, and it shows further still that many German designs and projects near the end of the war were more than simple fantasies or crude and desperate attempts at an end product.

A swift, uncomplicated solution Project P1095: After Hans Hornung and his group, in connection with the work on the P1092, came upon the development of the Me 262 for the multi-purpose fighter aircraft, another project team from the Messerschmitt group was working independently from him on a single-engine fighter aircraft. Since 1941, Engineers Prager and Mende and their group leader, Engineer Seitz dealt first in Augsburg and later in Darmstadt with a new type of aircraft which was almost ready for mass production in the second half of 1943.

Even with the first achievements and studies which emerged from the 1079 project for the Me 328 aircraft, other functions and uses were anticipated for the light fighter. In October 1943, the German Air Ministry (RLM) discontinued the long-range attack bomber, Me 328B, due to engine problems for the benefit of the War Cabinet.

Seitz, in full agreement with his co-workers, knew that under no circumstances should they let the fully developed aircraft quietly fall to the wayside. Hence, prompted by the Argus exhaust system, he undertook the experiment to equip the Me 328 with the Jumo 004 engine. He had hoped to be able to offer a fighter aircraft to the German Air Ministry which, because of production reasons, would not be equipped with the most modern aerodynamics, but which would be available, however, through its simple and robust type of construction quickly and in large numbers, and ready to defend bombers. A scant year later, the identical thoughts led the RLM to the Heinkel He 162.

The first calculations and designs put together by licensee Jacobs-Schweyer in Darmstadt for Prager concerning the Me 328 showed that the aim for everything else was elementary.

Aircraft Type	P1092 a	Arsenal VG70
Primary function	single-engine fighter	swept-wing research aircraft
Crew	1 pilot	1 pilot
Power plant	1 x Jumo 004C with 1015 kp thrust	1 x Jumo 004-B2 with 890 kp thrust
Length	8100 mm	9700 mm
Wing span	7750 mm	8500 mm
Wing surface	12.7 m2	15 m2
Aspect ratio	4.75	4.8
Sweep	20 degrees	35 degrees
Average profile	0011.3	
Take-off weight	3664 kg	3393 kg
Wing load	288.5 kg/m2	226 kg/m2
Maximum speed	925 km/h at altitude of 8 km	900 km/h at altitude of 7 km
Service ceiling	11,600 meters	10.670 meters

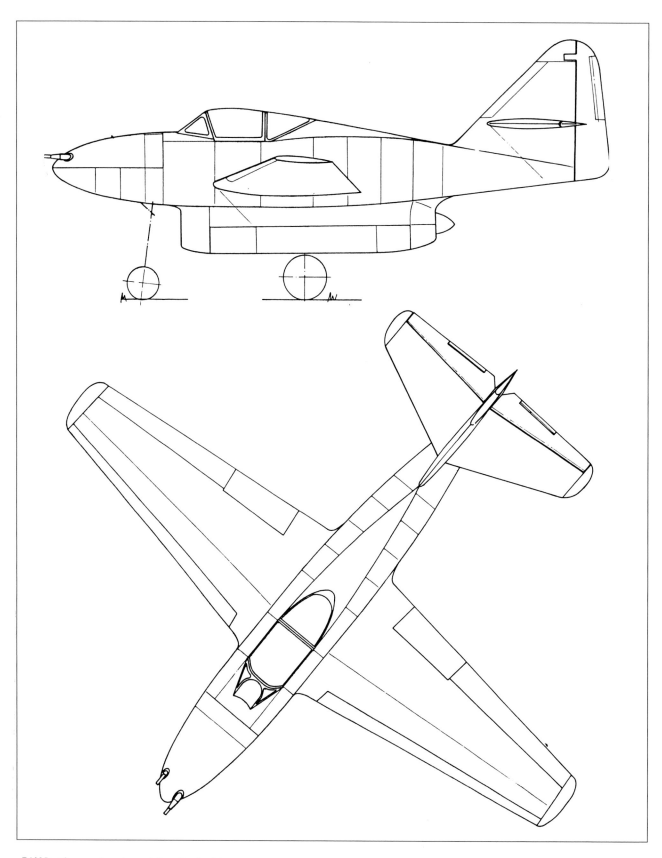

P1095 with a wooden wing and the tail unit of the
Me-262.

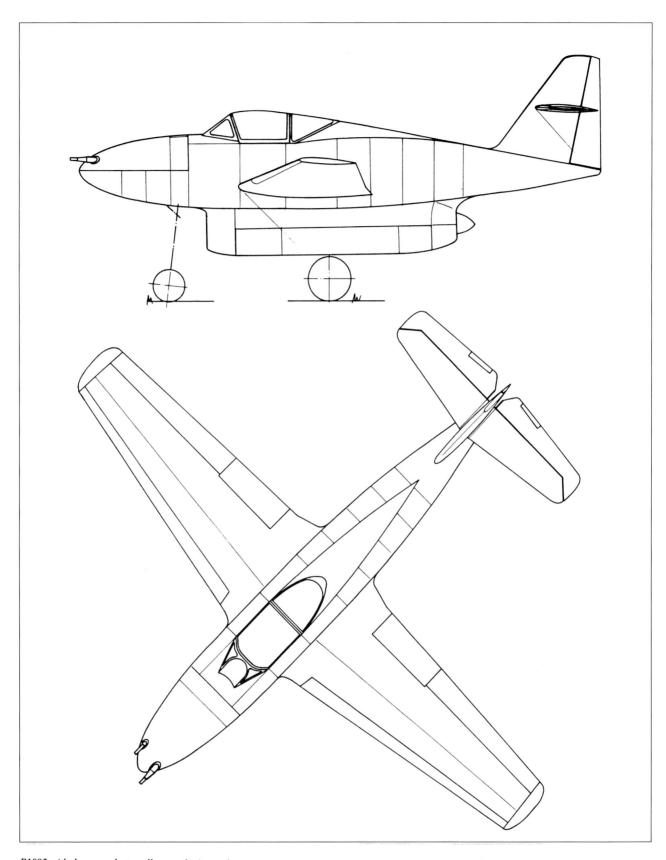

*P1095 with the somewhat smaller metal wing and
the tail unit of the Me-328.*

The idea didn't even survive the first discussions with Professor Messerschmitt in Augsburg; the determining factor was the extremely high wing load of the Me 328, which could reach over 500 kg/m2.

The change from the extremely light Argus intermittent air jet (IL) engine to the substantially heavier Junkers 004-engine would have driven up the value to the extent that it would have made the aircraft unacceptable for load capacity.

Following this slight modification, the turbo-jet driven Me 328, hot off the press, (and quite often falsely labeled the Me 328C), was almost completely modified by Seitz. The aircraft's basis for design was to use, to the greatest extent possible, the already existing subassemblies.

The following short description of the project illustrates the "collage effect" of the aircraft's structural approach, which would guarantee a quick acceptance for mass production without all the developmental risks.

Primary Function:
Single-engine, single-seat fighter aircraft with sturdy construction

Engine:
1 x Jumo 109-004 B with approx. 900 kp static
    thrust

Fuselage:
Total length 8400 mm with tail unit Me 262
Total length 8050 mm with tail unit Me 328

The fuselage nose shows construction of a "weapons turret." Both of the fixed Mk108's are mounted in the same manner as in the Bf 110G-4/R3 night-time fighter.

The test-bed Me-328, the likeness to the P1095 is easily visible in this photograph.

The retractable nose wheel has a size of 465 x 165 mm. The design called for the landing-gear strut to be located directly in front of the air intake. Junkers tested the strut in Dessau on October 13, 1943, for loss of thrust caused by the strut.

A 200 liter fuel tank is housed between the nose and cockpit and two additional tanks with 590 liter capabilities are located directly behind the pilot. The space between the fuel tanks and the tail unit is provided for housing (FT) communications equipment.

The engine is easily serviced, i.e. readily accessible with bolts fastened to the fuselage's support structure.

Wings:
For the wing assembly, Seitz proposed the typical Messerschmitt shape which could be made from metal or wood. Both versions show the entire wing span's leading-edge flaps, tail unit of the large wing span and landing flaps. Directly behind the main spar, the retractable main undercarriage is fastened to the inside of the wing and has tires sized of 660 x 190 mm.

Wooden wing:
Wing span 9600 mm,
Wing area approximately 15 m2

Metal wing:
Wing span 9600 mm,
Wing area approximately 13,5 m^2

Selection of the undercarriage with a safe wheel load of 4600 kg brings one to the conclusion that the aircraft's take-off weight would be between 3500 and 4000 kg.

The reasons for discontinuing the project could be found in the inadequate technical concept. On one hand, the configuration of the air intakes at unfortified locations could cause engine damage; and on the other hand, the danger also existed that the adversary could be quick to match, performance wise, the aerodynamically conventional P1095.

Moreover, Seitz' group had sufficient work ahead with the Me 328, despite the non-acceptance as a "deep attack bomber." Beginning in 1944, production of further aircraft continued and those in charge discussed the employment the Me 328 as a "sacrificial" aircraft which should begin operation as early as the summer of 1944. Luckily, it remained a part of the discussion.

A New Generation is Born

The P1101 Experimentation and Fighter Aircraft

There are a whole series of well-known designs to be found under Project Number 1101, from which only a few have anything to do with the single-engine P1101. Therefore, a clarification is given:

To carry on combat aircraft projects P1099 and P1100, a series of studies for high-speed combat and bomber aircraft arose sometime between May and the middle of July, 1944. The designs had some common features in using the most modern aerodynamics of the day, together with the jet propulsion and proven Messerschmitt construction principles.

(Exception: the tailless design 1101/97 from May 22, 1944 was prompted by a Jumo 222 with pusher-type propeller.) Many of the solutions hinted at during this time are still encountered again and again in the years following the war. For example:

• the normal swept wing with 35 degree and 40 degree sweep, where the engine is installed in a gondola under the wing;

• swept wing with decreasing sweep to the outside ("crescent wing"). Above all others, this design for swept wing assembly was of great interest to the British firm Handley Page. The research aircraft HP 88 and the mass produced bomber and tanker aircraft HP 80 "Victor" were equipped with this form of wing, and were also based on the works of the Arado Firm and the Braunschweig Aviation Research Organization (LFA).

• a vertical take-off and landing (VTOL) design with swept surfaces and a main rotor of the Focke type. Following the war, Professor Messerschmitt seized this idea once again with the Me P.408 project "Rotorjet", and the relevance of this concept has lost

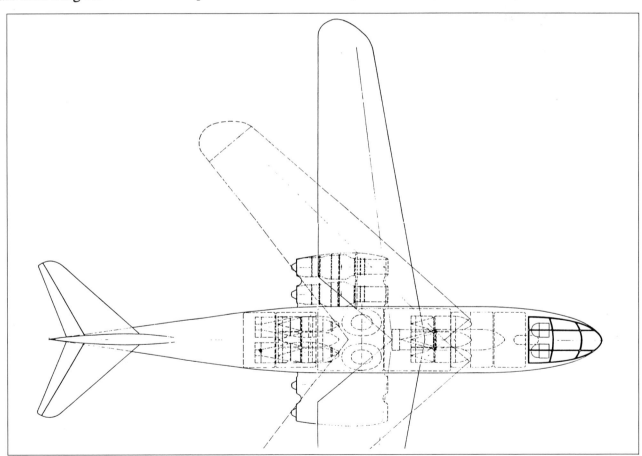

An example from the multitude of P1101 combat and bomber aircraft studies: design XVIII-103 from June 29. 1944; tile-wing aircraft with 4 x He S 011 engines (5200 kp total thrust); sweep variable between +8 degrees and +47 degrees; maximum speed approximately 1070 km/h (theoretically); wing span 14 m (47 degrees)/20.5 m (8 degrees) bomb load 6 x SC 500 or 1 x SC 1800/2 x SC 1000.

nothing even today;

• the variable geometric wing in split and unsplit variants

Messerschmitt's solution for the split tilt wing called for an arrangement of the center-of-gravity displacement for the forward and return movement of the complete wing assembly while in flight.

In the fifties, this solution experienced a virtual transformation in the American Bell X-5 and the Navy-counterpart, the Grumman XF 10 F-1 "Jaguar." The unsplit tiltable wing proposed by Dr. Richard Vogt, Chief Designer of Blohm and Voss Aircraft Construction, inspired Messerschmitt's project leaders to fantastic ideas. In several studies they brought aircraft designs to the drawing board which show "drift" or "rotary" wings in single- and double decker configurations (Design XVIII/108 from July 11, 1944).

The complete, thorough works and the practical realization of these studies still would have required quite a reach to prevent World War II from drawing to a close. Nevertheless, this "paper flyer" shows the beginning of a new chapter in aeronautical engineering and a wealth of practical paths and solutions to problems. And all of these proposals have one thing in common: despite having project number 1101, they have absolutely nothing to do with the single-engine fighter and experimental aircraft!

This confusion is only found by explaining figure comparisons, since incorporating the emergency fighter program and the resultant downgrading of the bomber program brought to an abrupt end studies on the high-speed combat and bomber aircraft, thereby making the no longer necessary project number 1101 again available.

On one hand, the emergency fighter program was the end of the multi-engine 1101 and on the other hand, it marks the beginning of development of the forward-looking, single-engine P1101.

In the summer 1944, the German leadership finally reacted to the catastrophic, raw material shortage and the emergence of the allies long-range fighter planes, which the English and Americans used for protection during bombing raids over the German Empire.

An effective defense for the heavily armed bomber became a more and more unsolvable task for the defenders. The German Luftwaffe could still reverse the process only through a massive number of aircraft and/or through clear technical superiority. Since even an approximate numerical balance would still be considered wishful thinking, technical superiority remained a wafer-thin chance.

In the framework of the ongoing emergency fighter program, as ordered by the highest leaders, the already perceived drawback and tasking of the bomber program (Heinkel He 177) resulted in a simultaneous increase in fighter production (and mostly the reciprocating engine fighter, the Bf 109!).

The only available aircraft test-bed however, which offered the sufficient technical advantage in this situation, was the twin-engine Messerschmitt Me 262, which had been conceived at the beginning of the war in a totally different military and economical situation.

The aviation industry and some farsighted staff members of the RLM had evidently already recognized the raw material problem and had, for a long time, strove for a single-engine fighter, which was predominantly designed after the latest results of aerodynamic

research in order to offer the Luftwaffe an alternative to the Me 262. Some examples are: the works of Blohm and Voss (B. V. P.197/P.198), Messerschmitt (P1092), Focke-Wulf (FW P.I-P. V/P. VII "Flitzer") and also the forward-looking designs of Dr. Lippisch in Augsburg for Messerschmitt and later at the Aviation Research Organization in Vienna.

Additionally, some experts expressed fear that the enemy could soon overtake the Me 262 technically.

From a Heinkel document of July 10, 1944 (immediately before the publication announcing development of a single-engine fighter): "Should a hostile single-seat jet be put into action, the Me 262, with its superiority, should not be counted on in all probability, due to its traditional construction type of unswept wings and a design with gondola-type engines on the wing (Note: this point of view refers to application of area rule) and its large resistances. It appears necessary therefore, to design and construct a faster aircraft with low construction costs." And further: "Only a considerable improvement of the critical speed compared to the Me 262 justifies a new design."

Although only a short time later Heinkel himself ignored this knowledge with the He 162 concept, they were not to be denied further. With a technically superior machine, half-way sufficient in number to engage enemy fighters and bombers, it still would have been pure suicide for the fighter aircraft.

A word on the Me 262: With its superior speed, its climbing ability, and its heavy weaponry, it was the ideal aircraft for battling the bomber groups. In

During the war he was the leader of the initial projects and after the war once again with Messerschmitt AG: Hans Hornung, here with Willy Radinger at the beginning of the 1960s while at a "trial seating" in a mock-up of the Me 308 "Jet-Typhoon."

the fight against allied escort fighters, curtailments had to be made: in maneuverability it was superior. Therefore, the only strict instruction issued to the pilots was not to engage in "dog fights." This disadvantage was not completely compensated for through improved weapons (for example: air-to-air rockets with a semi-automatic homing device and other similar equipment).

The order at that time could therefore only read:

• a single-engine fighter plane of the simplest construction type while staying away, to the greatest extent possible, from materiel shortages;

• with excellent flight performance, good maneuverability, that is to say — application of the most modern aerodynamics;

• with reasonable take-off and landing characteristics (requires only the shortest possible take-off and landing runways);

• supplied with modern, powerful weaponry and the newest electronics;

• and with a justifiable expense (for the time) for air-corps ground organization.

And those were the crude terms with which the Technical Bureau of the RLM requested, through Proposal 226/II on July 15, 1944, from the industry.

The RLM specified, in detail, the application of the Heinkel 109-011 A engine; furthermore it demanded a speed of roughly 1000 km/h at an altitude of 7 km, a fuel capacity of 830 kg (approx. 1000 liters), which would have sufficed for a flying time of about one-half hour (computed at ground level), and protection for the pilot against direct shelling with SmK ammunition 0.5" (12.7 mm) from the front, a

pressurized cockpit and the standard equipment for fighter aircraft.

In August the Technical Bureau expanded these standards. The Messerschmitt AG should consider the following for their design:

(TJ = turbo-jet engine, R = rockets)

1. Smallest TJ fighter - single-seater with the highest possible flight characteristics.

2. TJR fighter with rocket fuel for the fastest possible climbing or as an aid in battle, whereby the possibility of expansion through application of a fuselage connection should be considered.

3. The possibility of expanding the aircraft to a two-man crew employing a fuselage connection, whereby the space for the second man, additional weapons, additional fuel or an additional engine would be taken advantage of.

The other authorized firms, at this point in time, likewise in development stages, received similar specifications for guidelines pertaining to their designs (Heinkel and Focke-Wulf, and only later were Junkers, Blohm, and Voss included; Henschel worked on his own in early 1945 without requirements or guidelines).

Following is how Focke-Wulf described the specifications:

1. A TJ fighter with additional rocket engine for faster climbing at high altitudes.

2. A TJ fighter with additional rocket engine, but less rocket fuel, sufficient for high-speed climbing at medium altitudes.

3. A pure TJ fighter without rocket engine, with approximately 2 hours flying time at altitude of approximately 10-11 km.

Working in Bad Eilsen since December 1943, with the initial concept and development of the "Flitzer" production series were Tank, Multhopp, and Mittelhuber, among others, and Focke-Wulf only partially fulfilled these standards; mostly, criticism of the insufficient maximum speed was stressed, and this led to the totally new Ta 183 design.

The Heinkel firm was supposed to study the design "pure turbo-jet fighter." The famous Heinkel Project Leader, Siegfried Günther, could only prevail, however, with his design for a "Peoples' Fighter"; the proposals derived from the P1073 and the He 162 and the later flying wing design He P1078 B and C were of almost no use for this.

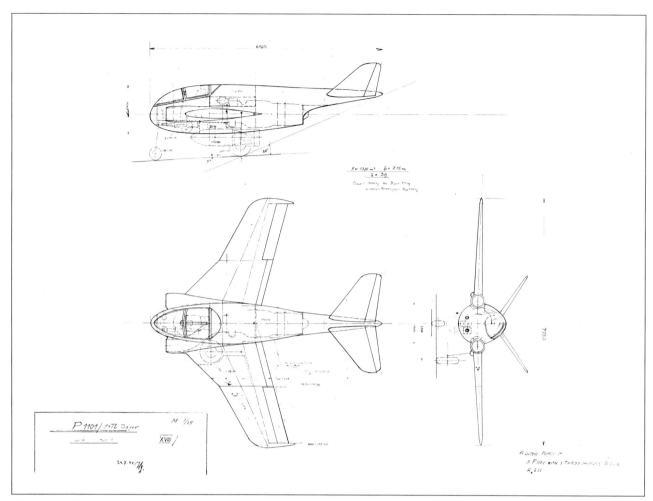

First design of the P1101 fighter from July 24, 1944.

Apparently the unusual interpretation of both of the last-named projects deterred the RLM's decision making.

Therefore, on 24 July, 1944, only nine days after announcement of the specification, the leader of Project Group 1, Engineer Hans Hornung, endorsed a study of the single-engine fighter aircraft under Project Number 1101. On paper, the aircraft clearly emerged from the experiences with the works of the P1092, but it has no similarity with the later construction.

The generously glass-enclosed cockpit required increased expenditure to protect the pilot.

The nose wheel should retract to the rear between both MK108's housed under the pilot. On the right and left of the fuselage nose Hornung ordered the circular air intakes for the engine located in the fuselage's middle. Above and below the Heinkel turbine engine,

710 liters of fuel can be contained. The semi tunnel-shaped engine exhaust required use of a steel plate on the underside of the circular tail boom, in which the communications equipment was housed.

In the foreground of the wing's configuration, the wish was unquestionably to eliminate the known disadvantages of the swept wing.

There was opposition to using the variable geometric winged aircraft for a possible solution to the substantial disadvantages of an aircraft of this magnitude and specification, both mechanically and weightwise. Installation of the tiltable wing promised decidedly greater advantages only for combat and bomber aircraft. Further development of aerodynamic technology fundamentally confirmed the outlook of Willy Messerschmitt.

The conception of the wing assembly showed a wing decreasingly swept to the outside without a slotted wing or leading edge flaps. To improve the slow-flying characteristics there were extension flaps of equal depth over the entire trailing edge flaps. When in the retracted position, flaps increase the profile camber and the wing surface and therefore improve lift as well.

The slighter sweep of the external wing section should effect better lateral stability at a larger angle of incidence; the greater sweep of the center section of the wing decreased pressure shock near the fuselage and allowed retraction of the main landing gear, through the large wing chord, into the area of the wing's root. Likewise, in the center section of the wing there were two fuel tanks with a capacity of 170 liters each. The total fuel capacity amounted to 1050 liters.

It is quite obvious that the first aircraft designed on paper would not yet satisfy requirements in many respects.

As has already been implied, the Technical Bureau called for Messerschmitt to enlarge the fuselage, if possible, and improved maneuverability of installed weapons was also a high point. Therefore, a series of designs and studies emerged in mid-summer 1944 to find the optimal solution for the task under the given requirements.

Two circumstances simultaneously hindered and accelerated the development of the end of the war:

• on one hand, the quickly changing military and economic situation of the German Empire;
• and on the other hand, the quickly increasing knowledge of the aerodynamic correlations closing in on flight at the speed of sound.

This were surely the main reasons for the emergence of a large number of projects, as was the case with German planning boards in the last years of the war.

A further study from August 22, 1944, in which a co-worker of Hornung, Project Engineer Thieme, showed something of quite another picture.

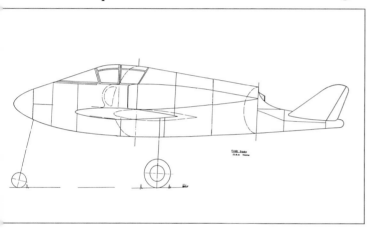

The P1101 study from August 22, 1944.

Technical data: P 1101 - Design from July 24, 1944

Main dimensions:

Total length	6850 mm	Landing gear:	
Height	2450 mm	Nose wheel	380 x 150 mm
Max. fuselage height	1400 mm	Main wheel	660 x 190 mm
Max. fuselage width	1200 mm	Undercarriage wheel track	2100 mm
Wing:		Maximum allowable load	4300 kg
Wing span	7150 mm		
Aspect ratio	= 3.9	Engine:	1 x Heinkel HeS 0011 with
Dihedral angle	2 degrees		1300 kp static thrust
Center wing section:		Weight:	
Wing span	3580 mm	Fuel	approx. 800 kg
Sweep (t/4-line)	40 degrees	Weapons and armor plating	approx. 400 kg
Profile	NACA 0010-0825-40	Take-off weight	approx. 3000 kg
	(NACA-National Advisory		
	Committee	Calculated flight characteristics:	
	for Aeronautics)	(Performance data)	
External wing section:		Maximum speed at altitude of 6 km 1050 km/h	
Sweep (t/4-line)	26 degrees	Climbing speed at sea level 26.8 m/sec	
Profile	NACA 0008-0825-40	Service ceiling	12,000 m
Tail assembly:			
	V-shaped tail with	Military Armament:	2 x Mk 108 release bombs
	110 degree angle		1 x SC 500

The aircraft had a two-person cockpit laid out. The engine exhaust lay over the tail cone which, as before, had a V-shaped tail unit.

One interesting point is the fuselage nose being designed as a weapons turret. Next to the Mk112, which would shoot in the direction of flight, there was an alternative vertical weapon designed for the aircraft, consisting of 2 x Mk 108 or a SG 500 "Fighting Fist."

The wing would bring one to the conclusion that a crescent wing would have been employed. And this would have meant a complete redesign of the wing assembly with the accompanying consequences in development, construction, testing, and production! Time was short and the war industry and officials were forced to return to the half-proven design and were to engage only in extreme circumstances in the risk to complete the new development.

At about this same time Messerschmitt hoped the practical experiments of the swept wing could soon begin with a similarly equipped Me 262. The test flights of the Me 262 HGII (Hochgeschwindigkeit-high-speed) aircraft at the beginning of 1945 already failed before the first flight, due to a ground mishap of the Me 262 W-Number 111538 equipped with a 35 degree swept wing.

It would suggest therefore, that the proven, mass-produced wing assembly of the Me 262, in a modified form, would be used for the new fighter aircraft.

During these last few days of the month of August, the Project Bureau in Oberammergau completed a P1101 design in the form prescribed by the Technical Bureau of the RLM ("Preliminary technical guidelines for a high-speed fighter aircraft with a jet engine" from January 4, 1939).

The design is endorsed by Hans Hornung, and it shows a stylish aircraft that reminds one of a modern jet trainer.

For the proposed aircraft, the project leaders also attempted to optimally fulfill the tasked requirements with the smallest possible surface areas. The reduced aerodynamic drag achieved through this process is one advantage, and a further advantage is that such an aircraft presents a target that's much more difficult for an adversary to hit.

Description of the P1101 design from August 30, 1944:

Fuselage
For attainment of the most slender-shaped fuselage for subsonic flight which, in contrast to the wing assembly, later received a totally different design, it is divided into four sections:

• The fuselage nose is modular as a "weapons turret" for different weapons variations.

To a large extent it is designed for a 55 mm Mk112 of Rhein metal and accordingly offers sufficient room for the alternatives, 2 x 30 mm Mk 108 or 2 x 30 mm Mk 103.

Weapons turret from the study from August 22, 1944. (below)

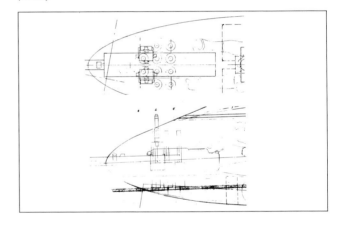

From the Messerschmitt Laboratory, a model of the Me-262 HG III (High-Speed) with 45 degree swept wing and 2 x He S 011 for flights near Mach 1. (on the right)

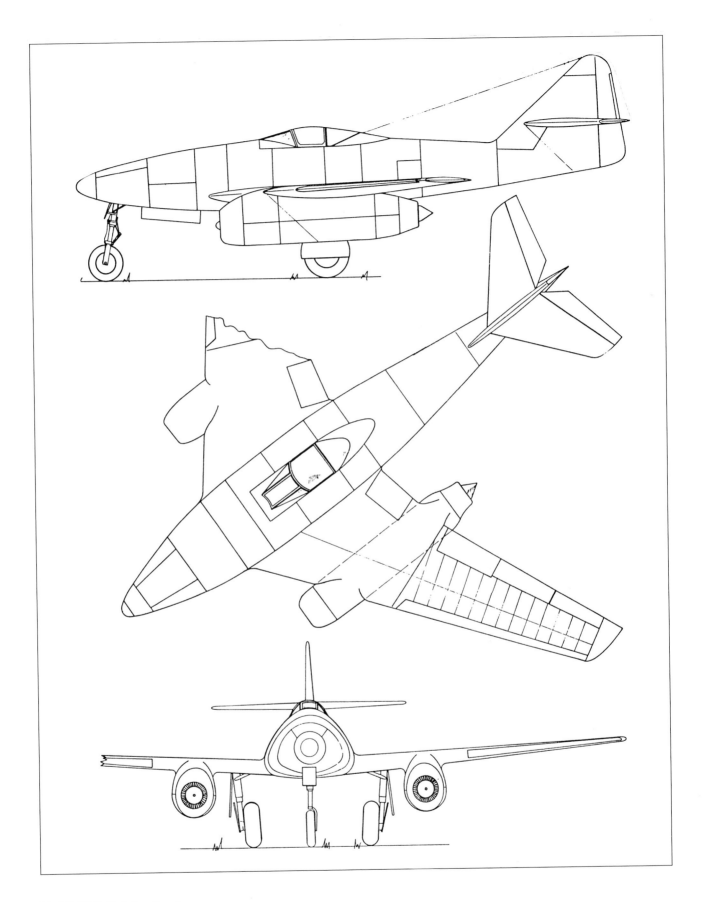

Me-262 HG II (High-Speed) with component test-ing (for example: wings) would have been possible for the P1101. Engine: 2 x Jumo 004 C.

In addition, in the nose section, the nose wheel is designed to retract in a rearward movement.

• Directly behind it, separated by a frame, the expandable cockpit is located with a connecting piece. The instruments and radio equipment for the cockpit match those of the Me 262 to a large degree. The pilot is protected from the front against shelling by 0.5 inch munitions by a thick, bullet-proof windshield. This protective design was further developed by installing a second bullet-proof windshield for mass production of the Me 163. The pilot literally sits upon 230 liters of jet fuel "J2"; fuel from this tank is used for engine warm up, engine start, and subsequent acceleration.

• On the left and right of the cockpit are circular air intakes (one on each side) for the Heinkel-Hirth HeS 011 jet engine installed in the fuselage's center section. Above the engine and behind the cockpit are the armored main fuel tanks. The total fuel capacity amounts to 830 kg. Also above the engine, behind the fuel tanks, the radio equipment is housed.

• The rear fuselage section is formed through a cone with a swept, V-shaped tail unit. Knowledge of a V-shaped tail unit had already been collected with a similarly equipped Bf 109F-4/V (VJ-WC Production No.14003).

Wing assembly

Aerodynamically, the design distinguishes itself mostly through its lifting surface with a wing sweep of 40 degrees in 25% wing chord (also 0.25 t or t/4 line).

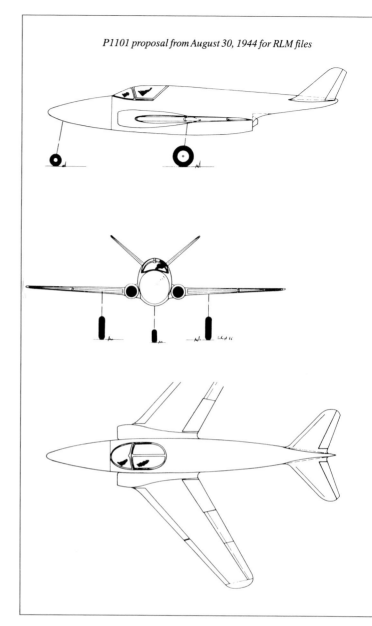

P1101 proposal from August 30, 1944 for RLM files

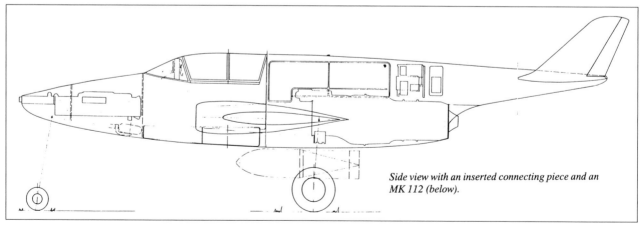

Side view with an inserted connecting piece and an MK 112 (below).

This lifting surface, in its basic characteristics, was replaced by the outer wing of the Me 262, possesses a divided leading edge flap that almost reaches over the entire wing span, a coupled tail-unit, and landing flaps.

The relative profile thickness decreases from 11% at the inner wing to 9% at the outer wing area. The profile originates from the American National Advisory Committee on Aeronautics series.

Undercarriage

The main undercarriage has a "pilot friendly" width of 3340 mm; it's to retract into the inner wing/fuselage area. The size of the main undercarriage tires is at 710 x 185 mm. As was already noted, the nose wheel (380 x 150 mm) retracts in a rearward movement into the weapons turret. When retracted, the nose wheel is tilted to a 90 degree angle.

Technical data on the P1101 design for a turbo-jet combat fighter from August 30, 1944:

Length:
without the fuselage connecting piece 9,370 mm
with fuselage connecting piece ... 10,020 mm
Height: .... 3,080 mm
Total surface area: .... 54,7 m²

Wing data:
Wing span8, ... 160 mm
Aspect ratio (b2/F) 4.94
Wing sweep .... 40 degrees in 25% wing chord
Wing surface 13.5m²

Engine:
1 x Heinkel-Hirth HeS 011 jet engine with a static thrust of 1300 kp; addition of booster rocket foreseen. The design also allowed for installation of a Junkers Jumo 004C.

Above all, this alternative was not to be sneezed at in regard to developmental problems of the HeS 011!

Military equipment:
2 x 30 mm Mk 108 cannons with 60 rounds each; with the possibility of carrying SC 500 release bombs.

Fuel capacity:
Additional fuel:
Outer tank: 600 liters
Towed tank with V1 wing: 1,200 liters
Flight weight:
(without additional fuel) 3,554 kg

Calculated performances:
Range without additional tanks: 1,500 km
with outer tank: 2,200 km
with towed tank: 3,000 km

Maximum speeds at altitude of 6 km:
without additional tanks: approx. 1,050 km/h
with outer tank: approx. 960 km/h
with towed tank: approx. 890 km/h
Climbing time to 14,000 m: 15 min.

Meanwhile, major problems cropped up in development of the new Heinkel jet engine, which had fallen further and further behind its anticipated timetable. The project leaders did not expect the first initial series tests (Heinkel HeS 011 A-0) before the summer of 1945. To a large degree, the fate of the new fighter aircraft would hinge on this engine!

On the other hand, however, given the present military situation, time was a pressing factor and to make matters worse, the extremely intense situation with raw materials left one no room to play with.

Therefore a certain logic was obvious when the RLM, on September 8, 1944, conveyed a new announcement to the aviation industry. Therein the Technical Bureau, as a temporary solution, ordered a high-speed fighter of the simplest construction form, while attempting to avoid using rationed materials; the light

fighter should be powered by a BMW 003 turbine.

On September 15, 1944, the victor from this announcement was the Heinkel He 162, whose mass production still began before end of the war.

Apart from this and at the request of the RLM, the leaders of Heinkel, Focke-Wulf and Messerschmitt met on September 10 in Oberammergau, in order to find a technically sophisticated, extensive, and forward-looking solution for the future fighter of the Luftwaffe. (A project, which in view of post-war development could be quietly assumed and was obviously handled thusly).

Heinkel used a developed project, partly from the 1073, for the basic design of the He 162. In contrast to the somewhat smaller He 162, Heinkel used a 35 degree swept wing with a wing span of 8.0 meters and a wing area of 14 m2; for an empty weight Heinkel's engineers calculated 2674 kg, while the take-off weight should total 3604 kg.

The Focke-Wulf firm presented the aforementioned Project VI "Flitzer", which displayed a similarity with the de Havilland D.H. 100 "Vampire" (already operational at this time), while Messerschmitt submitted the already comprehensively discussed P1101 design from August 30, 1944.

This conference was to serve as a first discussion, a comparison of designs, and of computation procedures.

It showed that above all, weaponry was difficult to come to an agreement on; though an agreement was reached on operational matters, whereby 2 x Mk 108 and 60-round weapons were to be employed.

The RLM required the firms to provide an internal fuel capacity of 1200 kg in their designs. Instead of the up-to-now required one-half hour of full throttle flight at its operational altitude, it was to now be possible for full throttle flight for one full hour; most likely that the powers that be wanted to disassociate themselves from the "Peoples' Fighter" with this extended capability, which demanded full-throttle thrust for 30 minutes of flight time. (In early 1945, the RLM desired the inclusion of a further increase in the internal fuel tank capacity in the design.)

It was somewhat more difficult to work out flight performance calculations as individual companies had considerably different calculation procedures. This is thoroughly understandable however, as they planned nothing short of busting through the speed barrier up to the speed of sound!

Final agreement for a common calculation procedure, which is an unavoidable requirement for a (theoretical) comparison of designs, was to resolve itself during later meetings. These types of discussions, comparisons, cooperation, and this struggle for the optimal solution on a theoretical basis was totally new.

Up to this point in time each aircraft manufacturer would lay out his proposal with the usual observance of secrecy against the competition. The RLM's Technical Bureau provided only the aircraft's tasking and the design parameters and/or a somewhat narrow outline. The RLM, as far as it was concerned, had always had grief with the task of comparing the various documents submitted by the different firms, since no uniform procedural guidelines were in place.

The consequence of this was a comprehensive trial of the various test beds and, when available, the aircraft's first production series, which included one or several comparative flights and also an operational test. The finality of the tests was the RLM deciding a victor, and with it, the future prototype for the Luftwaffe.

As has already been noted, the urgently required answer to defend against allied bombing raids with emergency and interim solutions like the Heinkel He 162, with a manned anti-aircraft missile, as for example, the Bachem Ba 349 "Natter" (Viper) and finally with the so-called "miniature fighter proposal" from November 1944, at which even the standards for the "Peoples' Fighter" were once more simplified, mostly concerning the engine.

All of these attempts finally proved to be dead ends, and after the war, they had only episodic meaning for further development of military as well as civil aviation.

The Me 262 and its designs, by comparison to its successors, formed the basis for some the most successful post-war designs of the victorious powers. A further, more direct measure on the more organizational levels had triggered the theoretical difficulties of the Oberammergau meetings: the RLM and those heading the DVL (Testing Laboratory for Aviation) were evidently attempting to place the fundamental works of the common performance calculations on a broader basis. Therefore, two aviation firms (Junkers, Blohm and Voss) received development contracts from the RLM for the current bid. The two firms had not, up to this point, emerged in the area of aircraft construction, but had dealt for quite some time with research on high-speed flight.

For Messerschmitt the meeting was cause to publish guidelines for construction of future production aircraft. Structural to these guidelines (dated September 13, 1944) and under consideration of the latest knowledge from aerodynamic research, the Project Bureau pursued an improved, extensively new design for the P1101.

The aircraft, which was to later be built both in Germany and in the USA as an "Experimentation Aircraft", and which also stood as a basic model for a generation of jet fighters, was born in mid September, 1944! This production pace and the results achieved under the given conditions of this work are more than impressive and once again secure a special place in the history of aviation for Messerschmitt development!

Meanwhile, the swept wing research, which began in 1940, continued at full speed in the wind tunnels of the industrial complex and the research institutes.

The activities brought about by the Messerschmitt firm mainly encompassed three to six component tests and stress distribution/pressure distribution records in wind tunnels, which permitted air speeds up to the trans-sonic range. New, pioneering testing facilities were under construction or in the planning: such as: the planned supersonic wind tunnel from Dr. Peters in the Ötztal. (This facility was created in 1953 by the French ONERA (National Research and Study Office) in Modane-Arrieux).

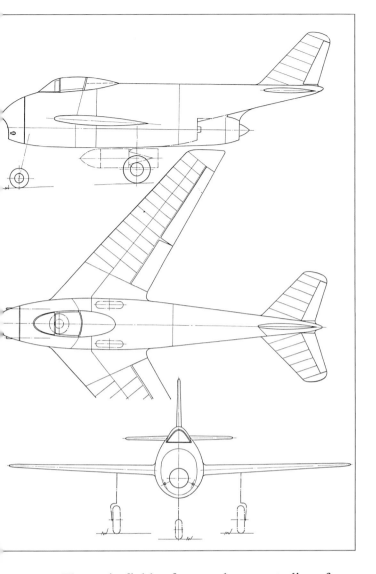

P1101 proposal October 3, 1944; geometric data could be found in the wind tunnel design from October 22, 1944.

The main fields of research were studies of:

• the most diversified swept wing forms right up to the delta wing;

• the fuselage and gondola influence on the wing assembly (tests conducted for area rule concept);

• and the optimal construction and configuration of the tail unit for wing assembly.

The size of most of the wing assemblies, half- and full assemblies, manufactured by the commissioned firms conformed to the size of the wind tunnels used. Testing took place firstly in the following facilities: the Göttingen Aerodynamics Testing Laboratory (Aerodynamische Versuchsanstalt)-AVA, the German Testing Laboratory for Aviation (DVL) in Berlin-Adlershof, the Aeronautic Research Institute (LFA) "Hermann Göring" in Braunschweig-Völkenrode, whose largest wind tunnel was also appropriately called "Hermann Göring" at the Aerodynamic Institute of the University of Braunschweig

(AIB), and last but not least, in Messerschmitt's very own wind tunnels which had been in operation since the beginning of 1943 in Augsburg-Haunstetten

Development suffered from the constant over-tasking of the research institutes due to the multiple requests by the industry; the beginning, fundamental work required careful coordination in order to prevent the overlap of similar tasks from the beginning.

The chords of the swept wing experiments at that time were tested by the "technical specialist for swept wings in a special wind tunnel"; the technical specialist was, since July 1943, Messerschmitt staff member, Engineer Puffert.

The somewhat lengthy testing method, the over tasking of the research facilities, and the impractical, unfulfillable observations which the technical innovations provided (as, for example: interaction between jet propulsion and the swept wing) led the forward-thinking Professor Messerschmitt to the thought, the decision, and the construction of a flying test bed which could simultaneously be used as a prototype for mass production.

By employing the already available components (wing assembly, undercarriage, engine, and control), on a large scale, the project had all the resources and the chance to be realized in a relatively short time. The decision to construct an "Experimentation Aircraft" came to Messerschmitt alone, and without guidance from the Technical Bureau of the RLM!

Aside from the circumstance of having a jump on the competition, which should not be underestimated, this procedure offered an advantage in that Professor Messerschmitt and his project team could corroborate their calculations and decide for themselves what was acceptable via practical results, as they were no longer dependent, to such a full extent, on the theoretical, comparative method, which had to first be established by the DVL (German Testing Laboratory for Aviation).

Through this point of view, one would think of the proposal from August 30, 1944. The result of this revision, adjustment, and further development was a design new in multiple respects.

A full view drawing from the Project Bureau showed an aircraft with all the characteristics of the post-war generation of jet fighters:

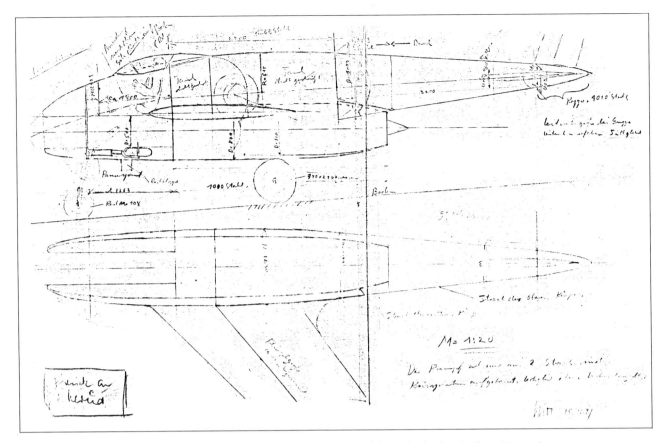

- air intake for the completely faired jet engine inside the fuselage nose;

- relatively sharp sweep (40 degrees on the t/4-line), thinner wing assembly, which presents a relatively large aspect ratio;

- swept vertical and horizontal tail in the standard construction;

- superimposed clear vision cockpit;

- and of course a tricycle undercarriage with a nose wheel.

The aircraft was specifically designed for the Heinkel HeS 011. For the Experimentation Aircraft, the Junkers Jumo 004B engine, with its diminished thrust and short-term availability, was to be used. Engine retooling required no great expense. This idea was a winner, considering the fact that the Heinkel engine was not in any sort of foreseeable production stage, as of yet. A short time later at Focke-Wulf, development with the Ta 183 went along the very same lines.

Hand-drawn sketches from Professor Messerschmitt showing the "Experimentation Aircraft" from October 15, 1944. The attempt for a tear-drop shaped, rotationally symmetrical fuselage form is clearly recognizable (above).

Vorläufiger Ausstattungsplan PL Serienausführung	Zelle u. Triebwerk	FT	Zielgerät	Abwurfvisier	Schußwaffe	Abwurfwaffe	Allgemein	Panze
A) Ständige Ausstattg. für alle Varianten	2 x 500 kg Starthilfen abwerfbar	FuG 15 FuG 25 a	—	—	—	Befestigungsmöglichkeit von Bombenschlössern an Rumpf und Fläche	Blindflugtafel Ia Fernkompaßanlage Triebwerksüberwachung Warmluftspülung der Sichtscheiben 2 x 2 Ltr. O₂ Flaschen 3000 Watt Generator Folien-Antenne auf Holz-Seitenleitwerk	keine scheib klappl platte
B) Spezielle Ausstattg. (zusätzlich zu A)								
I) Schönwetterjäger ev. aus Experimentierflugzeug (wird durch Schönwetterjäger II abgelöst)	Triebwerk Jumo 004		Revi 16 B		2 x MK 108 je 100 Schuß	Rüstsatz für Abwurfanlage Schloß 503 (als Rüstsatz)	Jägerkurssteuerung Schußkamera	von 12,7 (als
II) Schönwetterjäger (Hauptaufgabe I)	Triebwerk HeS 11 Druckkabine Abwurfbehälter für Kraftstoff	FuG 218 FuG 520 FuG 206	EZ 42	TSA D2 (als Rüstsatz)	2 x MK 108 je 100 Schuß Raketengeschosse Großkal. 2 x AG 140 oder Kleinkal. 5 cm Batt. Schußwaffe mit Fotozelle 12 SG 116 oder 8 x 108 Batt. oder Vielfachwaffe	Rüstsatz für Abwurfanlage Schloß 503 (als Rüstsatz) 4 x X4 oder 2 x HS 298	Jägerkurssteuerung Schußkamera	von 12,7 (als
Schlechtwetterjäger (Hauptaufgabe II)	Triebwerk HeS 11 Druckkabine Abwurfbehälter für Kraftstoff	FuG 218 FuG 125 FuG 120 FuG 520 FuG 206	EZ 42	TSA D2 (als Rüstsatz)	2 x MK 108 je 100 Schuß Raketengeschosse s. o. Schußwaffe mit Fotozelle (siehe oben)	Rüstsatz für Abwurfanlage Schloß 503 (als Rüstsatz) 4 x X4 oder 2 x HS 298	Jägerkurssteuerung Schußkamera	von 12,7 wärts (als
Nachtjäger (einsitzig) „Wilde Sau"	Triebwerk HeS 11	FuG 218 FuG 125 FuG 120 FuG 520 FuG 101	EZ 42		2 x MK 108 je 100 Schuß Schrägeinbau 1 x MK 108 mit 150 Schuß		Jägerkurssteuerung Schußkamera	von 12,7
Interzeptor	Triebwerk HeS 011 Druckkabine Zusatzschubgerät	FuG 520 FuG 206	EZ 42		2 x MK 108 je 100 Schuß	Rüstsatz für Abwurfanlage Schloß 503 (als Rüstsatz) 4 x X4 oder 2 x HS 298	Schußkamera	von 12,7
Nahaufklärer	Triebwerk HeS 11 Druckkabine Abwurfbehälter für Kraftstoff	FuG	Revi 16 B		1 x MK 108, 100 Schuß		Robot RB 50/18	von gege

In a sketch that survived Professor Messerschmitt from October 15, 1944, he detailed his planned Experimentation Aircraft. Using a 1:20 scale, the sketch matches, to a large degree, the aircraft from Messerschmitt's workshops; the main undercarriage retracted inward to the front. The Project Bureau once more used this design in February, 1945, in a study of the production proposal, and Swedish designers at SAAB in Linköpping cleverly converted it into their proposal, the J-20 "Tunnan", which was greatly influenced by the P1101.

With this design, a more favorable load distribution was attempted to circumvent the somewhat difficult accommodation of the undercarriage in the area of the engine exhaust.

By the end of October, a working sketch of a wind tunnel model in the scale of 1:10 was produced. Where and when a model, manufactured according to this sketch and used for testing and analysis, would be produced would doubtless be clarified. In all probability the AVA conducted tests for the high-set, horizontal tail unit in Göttingen beginning on November 15, 1944. At the beginning of November the Project Bureau prepared the "preliminary configuration plan", which represents the P1101 production series in five different equipment conditions set apart for the different missions. (see table)

And shortly thereafter the leader of Project Group 1, Engineer Hans Hornung brought the preparation tests on the experimentation aircraft to a provisional conclusion: on November 10, the "official" project delivery ensued. This marked the handover of design documents and construction data from the Project Bureau and Construction Bureau.

The data from the individual departments contained a design for the entire aircraft, an exact weight distribution (= requirement), and a description of the components with a pertinent full-view sketch.

Since the planned experimentation aircraft could also be used as a prototype for the production model, requirements in the delivery protocol were taken into consideration for standard production.

P1101 wind tunnel model from October 22, 1944.

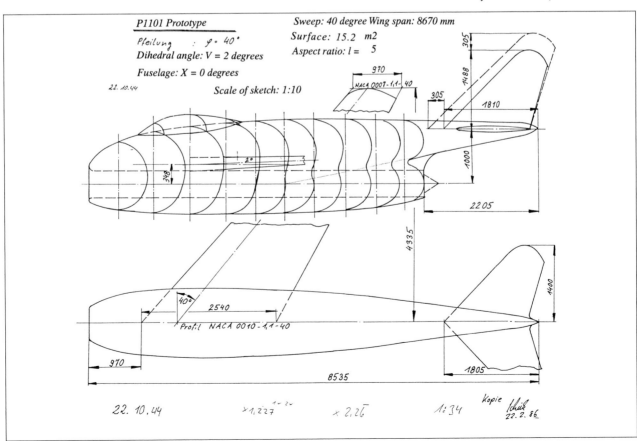

Secret!

Project delivery   P1101

Page 1

1st Section: In General
A) Project guidelines:

The P 1101 project is designed as a single-seat fighter aircraft with a turbo-jet engine and with the He S 011 engine as the final choice.

An overview of the planned development stages is provided in the preliminary configuration plan from November 2, 1944.

In order to test the aircraft as quickly as possible, and to accelerate the later mass production, the following requirement is levied: use as many components and sub-assemblies as possible from the mass-produced Me 262, without modifications.

The first aircraft will be built as an experimentation aircraft, that is to say, it will be able to undergo modifications of the wing, fuselage and tail unit during testing.

The experimental aircraft can become the prototype for the production series, but that is not absolute. i.e. a longer processing period might be necessary for thorough testing of the configuration proportion, the status of various equipment for the individual stages of development, and for testing flight characteristics of the highly swept wings and tail unit and their response at over-critical Mach numbers, and for studying the sturdy fuselage stages with jet effect.

The requirement for the production series, as far as they are already set up, must be considered in construction of the first high-speed aircraft.

For the moment, project delivery is concerned with the experimental aircraft with the known requirements for the production series. It will be continuously expanded through inclusion of results from testing the experimental aircraft and the completed tests for equipment status and its installation.

In order to obtain optimum performance, it is absolutely necessary that during construction and production, a maximum of aerodynamic surface finishes be employed. The surfaces employed for the aircraft up to this point are insufficient for reaching maximum speeds.

Oberammergau, November 10, 1944
(Skilled Specialist)
(Department Superintendent)
(Professor Messerschmitt)

Distribution:
KL
Prototype Director
Sta Bureau
Project Bureau

2nd Section: Experimentation Aircraft (Design XVIII/138)

Tasking for the experimental aircraft:

The Experimentation Aircraft is designed as a test aircraft for the highly-swept wing (without gondola effect on the wing), for take-off and landing as well as for maximum speeds and as a precursor for mass-production. The proportionate configuration, the fuselage fairing, the tail units and wings are to be designed only through calculated testing, and it is possible that exact analysis and test modifications will be necessary during transition to a production series.

A) Fuselage:

I. Fuselage fairing (Design Number: XVIII /131 a and b)

The fuselage is composed of an upper and lower body of revolution with a connecting piece that joins the two.

Design XVIII /131 a and b is to be used as a base scale for the fuselage fairing.

The greatest effort is to be placed on exact compliance of the nose shape in the air inlet at the fuselage nose.

For the time being, the fairing for the Jumo 109-004-B engine is to be utilized.

II. Fuselage breakdown (Design XVIII/130 for the production series and XVIII/138 for the experimentation aircraft)

1.) The fuselage nose section contains: the air duct tube for the engine, the crew, the armament, the armor-plating, the pilot's cockpit canopy and the nose-wheel undercarriage.

The following points are to be considered in reference to the later dihedral mock-up aircraft by constructing the fuselage nose section for the experimental aircraft:

The 500 x 180 wheel easily housed.

The weapons mounting area must also be spacious.

The armor-plating must be useable as standard equipment.

The cockpit must be able to be made pressurized.

The cabin will be constructed as a racing cockpit.

At the air duct tube, note that the inner wall must be constructed as smooth as possible.

The rivet heads and butt joints are not to be moveable.

"P1101 Project Delivery" document.

2.) The middle fuselage section contains the fuel tank in the upper portion and the retracted undercarriage is underneath the engine.

For the experimental aircraft the following is proposed:

Jumo 109 004 B engine. But, subsequent installation of the He S 11 must be possible. Protected fuel tanks are not intended, therefore the fuselage's middle section of the tank can become tightly riveted, so that 900 liters can be contained. For the tightly riveted fuselage section (tank), a manhole is proposed. On the first protected tank, FuG 16 communications package from the Me 262 is to be installed.

The area of the landing gear is to have the 740 x 210 wheel. For the time being, the 660 x 190 wheels are to be installed.

The larger undercarriage will be standard.

Be aware of engine accessibility and ease of interchangability.

The transition from the upper to the lower body of revolution terminates at the end of the engine in a vertical edge.

Take care for best engine area seal from the fuel tanks.

3.) The tail of the fuselage is formed as a cone, and contains the communications equipment, the oxygen system, and possibly the deceleration parachute, the control of heading system, and the master compass. The tail unit is inserted at the fuselage tail end as well.

Be aware of the following:

When starting the engine, a flame may emerge at the engine's end. Therefore, at the very least, cover the fuselage underside in the area of the engine with a steel plate. Optimally, the fuselage should be constructed from sheet steel. Equipment installed near the engine's end are to be protected against radiation of heat.

Project delivery   P1101

Page 4

For maintenance and servicing of the communications system, control of heading, and the oxygen system, an adequate fuselage section must be designed.

Avoid construction of steel in the area of the master compass in the fuselage's end.

Fasten the compass to the fuselage's end, so that only the hood the fuselage's end can be moved over it.

Oberammergau, November 11, 1944
(Skilled Specialist)
(Department Superintendent)
(Professor Messerschmitt)

Distribution:
KL
Prototype Director
Sta Bureau
Project Bureau

B) Wing assembly: (Design XVIII/134)

As a whole, the wing assembly is designed after the Me 262 wings. Rib 21 to approximately rib 7 are used to this end.

The wing tip is modified to 40 degrees and thereby enlarged.

The wing's initial setting is swept back to a setting of 40 degrees.

The aircraft's profiling, type of trailing edge flap and trailing edge profile section, aileron section and slot are to applied as with the Me 262.

The wing's angular setting to the engine's axle is equal to 0 degrees.

The wing is mounted into the fuselage but is still separable.

For the experimentation aircraft, the following modifications are to be accomplished on the wing section carried over from the Me 262:

The wing must be able to rotate to the point that a sweep of 35 and 45 can be accomplished and is possible to 0.25 l.

The dihedral angle must be adjustable between + 2 and -3 degrees.

Normal dihedral angle is to be 0 degrees.

The outer aileron is to be extended over the wing span of the wing tip.

The inner aileron opens and is replaced by the landing flap.

During testing, the aileron may turn out to be too small. Therefore, at the expense of the landing flap's span, enlargement of the inner wing span must be possible.

For the first experimentation aircraft, lengthening of the inner slots to the fuselage need not be used.

A rather high loss at camax (40% of the Me 262) is to be expected.

When lengthening the inner slots to the fuselage later, beware that the aerodynamic efficiency of the inner slots is worse than the outer slots in order to obtain safety against nose diving. This can be accomplished through

Secret!

Project Delivery P1101
Page 6

Larger depth of the outer slot or through modification of the inner slot's angle of extension to the outer slot at equal depths.

After wind tunnel testing, expect that at a 40 degrees wing sweep the slot depth with 13% (with the Me 262) will be too small and must be increased to 20%. Thus, it is necessary that a second wing section with deepened slots (to 20%) should be prepared for testing.

Beware of the wing's surface finish. The space between the aileron, landing flaps, and slots should be as small as possible and should lie in the direction of flight.

An internal aileron engine is required and a servo may be necessary.

There is no counter balance outside of the profile.

Exceeding the profile contours with equipment, engine fittings or anything else should be painstakingly avoided.

A compilation of the requirements in order to configure the best possible aerodynamic surface will be reached in due time.

Oberammergau, November 10, 1944
(Skilled Specialist)
(Department Superintendent)
(Professor Messerschmitt)

Distribution:
KL
Prototype Director
Sta Bureau
Project Bureau

C) Tail unit: (Design XVIII/135 a and b)

The vertical tail can only be applied after receiving the mock-up's measuring results. For preliminary dimensioning, the vertical tail surface is to be attached with 1.8 m^2.

The horizontal tail is newly constructed out of wood with a surface of SW = 2.5 m2. The surface must be increased through a larger wing tip edge up to SW = 2.8 m2. The

elevator is straight with 30% of the tail unit's thickness ratio to the unenlarged wing tip end. It has aerodynamic internal balance, whereby the rotation axis position can be adjustable between 20% and 25% of the control surface chord; a trimming tab is to be employed in the horizontal tail unit.

Elevator deflectors + 30 degrees; tail surface trimming (against wing chord) from +3 to -9 degrees. Normal setting is approximately -1.

An alternate solution to devise is a horizontal tail of equal form but with 9 degrees positive aspect ratio. Furthermore, V-shaped tail unit analysis is to be carried out. Tail unit sizes for this will be addressed later.

Optimum surface finish should be taken into account.

D) Control:

To the greatest extent possible, the control system from the Me 262 is to be used. The fighter control of heading is to be provided.

The trimming and tail surface adjustment is to be executed mechanically.

Adjustment times and other items will be supplied later.

Project Delivery   P 1101

E) Undercarriage:

Nose unit:

Use sturdy 500 x 180 wheels:

Install 465 x 165 wheels in the experimental aircraft.

The nose wheel is not braked.

The shock absorber strut can probably be replaced by the 410 skid.

The incline should be neither over 30% nor under 15%.

Due to the turning radius on the ground, the hydraulic flutter damper is to be used instead of a frictional flutter damper.

Data concerning measurements of the damper cylinder and the required traverse angle

are indicated separately.

Pay attention to easy rotation with low extension speed.

Main undercarriage:

Wheels: Use sturdy wheels (for the airframe) 740 x 210.

Install 660 x 190 wheels in the experimental aircraft.

The shock absorber strut from the Bf 109 can probably be used. Test the wheel strength.

Ground clearance should not be too low nor wheel base too narrow. Pay attention to sufficient torsional rigidity of the landing gear mount to obtain low torsional stress in modification of toe in modification at the brakes. The decline should be, by no means, larger than 2 degrees.

Keeping the ability to taxi in consideration, the main wheel should be mounted as close as possible to the center of gravity.

Oberammergau, November 10, 1944.
(Skilled Specialist)
(Department Superintendent)
(Professor Messerschmitt)

Distribution:
KL
Prototype Director
Sta Bureau
Project Bureau

Project Delivery

P 1101

Page 10

F) Engine:

The He S 011 engine will be installed.

In the experimental aircraft and the first production series, the Jumo 004 B turbo-jet engine will be installed. Reconstruction of the He S 001 must be ensured.

The turbo-jet engine air intakes on the inside must be as smooth as possible.

Guidelines for more exact construction of the fuselage nose at the air intake system as well as construction of the intakes themselves will be supplied later.

Pay attention to easy engine accessibility, cooling, and compartmentalization of the fuel tank area. More exact data on these will also be supplied later.

For the experimental aircraft, the fuselage section over the engine is to be tightly riveted for taking in fuel (about 900 liters).

Information on fuel extraction and the wiring scheme will be explained later.

For emergency landings, a skid is to be provided at the engine.

Oberammergau, November 10, 1944.
(Skilled Specialist)
(Department Superintendent)
(Professor Messerschmitt)

Distribution:
KL
Prototype Director
Sta Bureau
Project Bureau

Project Delivery

P 1101

Page 11

G) Equipment:

The equipment for the production series has not yet been determined. It is to be supplied later, along with the type of installation. For orientation, a preliminary and yet to be analyzed equipment plan will be supplied, as well as some temporary installation types in the survey diagram.

The equipment plan for the experimental aircraft will encompass the following:

Instrumental flight panel I a
Engine monitor
Remote indicating compass
Fighter aircraft control
FuG 16 Radio, (from the Me 262)
3000 watt engine generator
20-hour amp
Warm air circulation for the window shield (extraction from the engine)

A more exact equipment listing will be supplied later.

Oberammergau, November 10, 1944.
(Skilled Specialist)
(Department Superintendent)
(Professor Messerschmitt)

Distribution:
KL
Prototype Director
Sta Bureau
Project Bureau

Project Delivery

P 1101

II) Armament:

As standard weaponry, the 2 x MK 108 with 100 rounds each is to be employed in the production series aircraft.

For the experimental aircraft the armament is to be delayed. It will only be applicable in regard to construction of the production series, and later installation is possible.

Armor-plating is to be delayed for the experimental aircraft. The possibility of later installation should remain available.

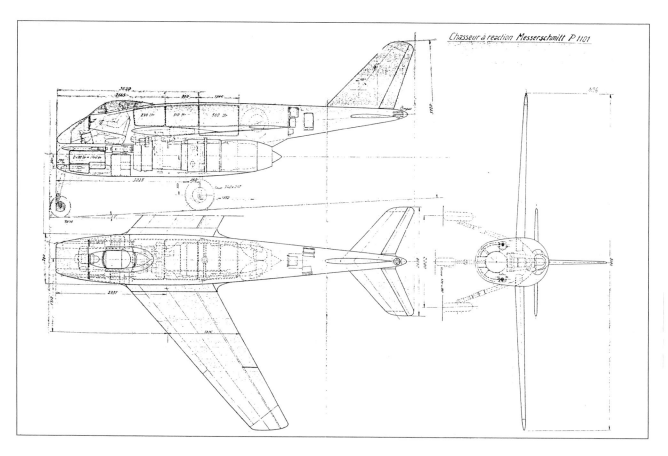

Chasseur à réaction Messerschmitt P 1101

Following project delivery the construction bureau staff members, whose leader since March 1938 was Walter Rethel, could begin establishing manufacturing work designations for the P1101-V1 experimental aircraft. Through interpretation and allocation of components, the construction office worked hand in hand with the manufacturing department (under the direction of Professor Julius Krauss); department head Georg Ebner was responsible for outlining the P1101 statics.

Since the selection of construction material for the P1101 is dated December 4, 1944, it can be assumed that experimental construction (under direction of

Moritz Asam) began during the first-half of December with the manufacturing of components.

The project group under Woldemar Voigt and Hans Hornung was amply busy at this point in time with supplying documents for the future fighter aircraft.

The standardized basis for computation had not been created, and also the concepts and requirements of the RLM's Technical Bureau were still anything but clear. Considering the oncoming critical situation, this is not remarkably odd.

Parallel to the 1101 production aircraft was a new proposal, developed from the 1101.

XVIII/37 Design from November 13, 1944. Overview of the production series aircraft (with Jumo 004 or He S 011 engine).

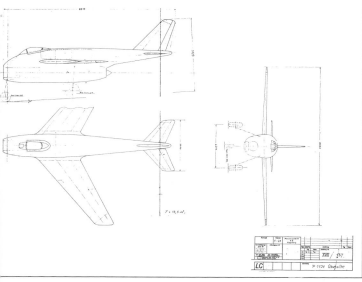

Detailed display of the production type. Design from a French archive; with "original" P1101 wing (in lieu of the Me 262 outer wing from the experimental aircraft). Indeed, the wing has nothing to do with the "A" wing assembly. The following details were planned in the fall of 1944:

- V-design with Jumo 004 B engine

- I. Day fighter with Jumo 004 B

- II. Day fighter with He S 011

- All-weather fighter with He S 011 engine and extensive electronic equipment.

For fighters, a take-off booster was to be provided, as well as a fighter-bomber conversion set.

- Reconnaissance aircraft with the He S 011, 2 x MK 108

- and finally for interceptor with He S 011 and the built-in communications equipment.

Here we have: The conception of a multi-role combat aircraft!

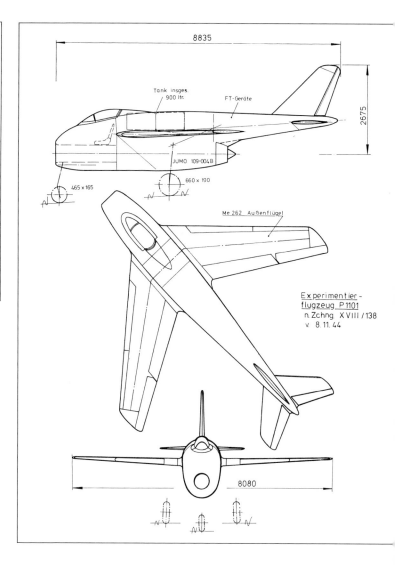

P1101 experimental aircraft according to Design XVIII/138. The difference from the production series is that this design called for less fuel and no military equipment (on the right).

This project, close to the current designs, received design number 1106.

On December 14, 1944, the project office presented the construction specifications of the single-seat, turbo-jet engine fighter P1101. The fighter described therein corresponds to the production series represented in the design from November 13.

P1101 specifications
(Single-seat turbo-jet engine fighter with HeS 011)
The aircraft is a single-seat fighter for high-speed flight with a turbo-jet engine.

Primary Functions:
Fighter
All-weather fighter
Fighter bomber
Interceptor

Total configuration:
Highly swept, mid-wing with retractable nose and main undercarriage.

P1101 production aircraft; with the He S 011 and 2 x MK 108 at the end of 1944. Constructed by: Günter Sengfelder (above).

Normal tail unit; in the final solution with a V-shaped tail unit. The turbo-jet engine (final solution HeS 011) lies under the wing in the lower fuselage half has an air intake tube extension in order to carry the combustion air from the nose to the engine.

The pilot sits in the fuselage section of the nose over the air intake tube in a pressurized cockpit.

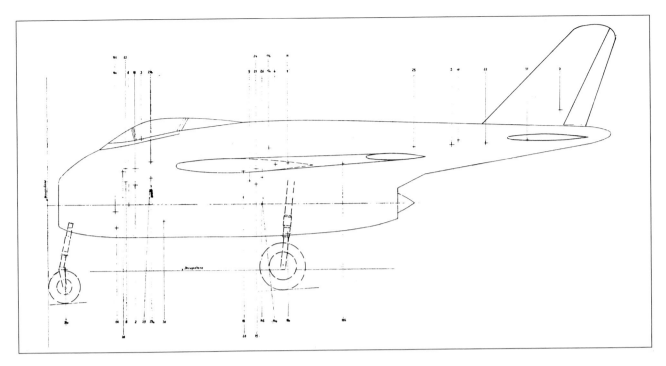

Behind the cockpit is an armored fuel tank for 1250 litters of fuel. Adjacent to the fuel tank lies the landing gear space for the main landing gear in retracted mode.

At the rear of the fuselage is the equipment for the inoperable instruments. Equipment such as with the normal fighter with fire extinguishing equipment, EZ 42 and mountable release and ejection system.

Nr.	Designation	Wt. (kg)
1	Wing	440.0
2	Fuselage nose	40.0
3	Cockpit + armor glass windscreen	50.0
4	Fuselage middle-section	85.0
5	Fuselage end	35.0
6	Air intake tube	80.0
7	Tail unit	65.0
8	Control: Fuselage nose	13.0
9	Control: Fuselage middle section	2.0
10	Control: Fuselage end	3.0
11	Control: Wings	20.0
12a	Main undercarriage (extended)	180.0
12b	Main undercarriage (retracted)	
13a	Nose wheel (extended)	50.0
13b	Nose wheel	
14a	Jumo 004 B engine	792.0
14b	He S11 engine	830.0
15	Auxiliary engine components	5.0
16a	Fuel tank	30.0
16b	Fuel	133.0
17a	Fuel tank	110.0
17b	Fuel	946.0
18	Hydraulics	25.0
19	Electrical components in the cockpit	30.0
20	Electrical components in fuselage mid section	18.0
21	Electrical components in the wings	5.0
22	Electrical components in fuselage end	11.0
23	Operational equipment in the cockpit	21.0
24	Operational equipment in fuselage and wings	12.0
25	Radio equipment	53.0
26	Fire-fighting equipment	27.0
27a	Cockpit	10.0
27b	Pilot	100.0
28	Armor-plating in the fuselage	120.0
29	Armament	190.0
30	Ammunition	117.0
31	Ballast	37.0

XVIII/133
December 12, 1944
Weight distribution of the preliminary P1101
production phase

Weaponry 2 x MK 108 with ~ 70 rounds.

Protection of the aircraft's pilot and the MK 108 ammunition through armor-plating.

On the wing and underside of the fuselage, fasteners for drop cargo release or additional fuel containers are to be provided.

Technical data of the P1101 proposal from 14 December 1944:

Primary function	single-engine light fighter
Crew	one pilot in a pressurized cockpit
Engine	1 x Heinkel HeS 011 with 1300 kp thrust and for take-off with maximum weight: additional rocket booster with 2 x 500 kp as intermediate solution or temporary solution: 1 x Junkers Jumo 004 B-2 with 890 kp thrust
Length	8915 mm
Height	3720 mm
Wing span	8080 mm
Wing surface	13.6 m2
Aspect ratio	= 4.8
Relative wing assembly thickness	12%-8.5%
Wing sweep t/4-line	40 degrees
Aerodynamic equipment	leading-edge flap over almost the entire wing surface
Undercarriage	
Nose wheel	500 x 180 mm
Main undercarriage	2 x 740 x 210 mm
Wheel base	2200 mm
Military equipment	2 x Mk108 each with 100 rounds and 500 kg release load

Weight	With the HeS 011	With the Jumo 004 B
Aerostructure	1008 kg	1008 kg
Engine	975 kg	937 kg
Construction weight	1983 kg	1945 kg
Standard equipment	159 kg	
Additional equipment	425 kg	
Empty weight	2567 kg	
Crew	100 kg	
Empty landing weight	2667 kg	
Additional load	1196 kg	

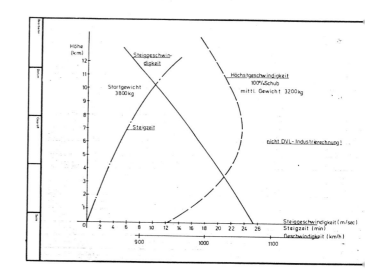

Take-off weight	3863 kg
Take-off weight with overload	4453 kg

(Take-off equipment and external stores)

Wing capacity:	normal	284 kg/m2
	maximum	327 kg/m2
Thrust load: (normal take-off)		2.97 kg/kp

For the technical leadership in Oberammergau the question was now, which of the two fighter proposals available up to this point should be presented to the RLM? In a theoretical consideration of total surface resistance, which has a decisive influence on the maximum speed, the P1106 comes away as the best:

	P1101	P1106	Me 262
Total resistance coefficient Cw	0.01978	0.01689	0.0227
Total surface	58.3 m^2	46.2 m^2	95.3 m^2
Engine thrust	1300 kp	1300 kp	1780 kp

The obviously higher aerodynamic quality from the table and, above all, the larger fuel capability of the P1106 tipped the decision in the end: after a detailed conference of the designs, Professor Messerschmitt decided to submit the P1106 proposal to the DVL in Berlin-Adlershof.

With the increased fuel capacity, Messerschmitt had responded to the demands expressed in late fall by the RLM's Technical Bureau. The planners in Berlin considered this indismissable, to clearly delimit the new fighter aircraft's profile against the interceptor in the style of the Me 163 or a fighter with a limited range, such as the Heinkel He 162. The 1101 project was not, as is often maintained in literature, refused by the RLM!

The mass production activities of the P1101 were discontinued and only in the second half of January, after the 1106 project was taken back, the activities resumed, to a certain extent.

The processes surrounding mass production had little influence on the works of the "Experimentation Aircraft."

The planned for and then completed aircraft was to be, first and foremost, a research aircraft for transsonic flight and a test-bed to undertake practical analysis of individual components of related designs.

The intent, round about the end of 1944 during the desolate time of decline, was to carry out a certain program and was created from the technical farsightedness of Professor Messerschmitt. Messerschmitt, who always put Messerschmitt the engineer before Messerschmitt the entrepreneur, was first and foremost interested in the continuing further development of aviation and its perfection.

Perhaps many of his designs and ideas were therefore resurrected from working so closely with and being subject to the directives of the Project Bureau, and the almost timeless validity of his works were proven. Certainly this is accurate for the works from the last months and days of the Third Reich in particular.

For its planned purpose, the conception of the "Experimentation Aircraft" was purely ideal:

The structural construction was simply maintained, and as far as it appeared justifiable, parts from mass-produced aircraft were used; above all, from the Me 262. Extensive changes to the main framework of the aircraft were possible without stretching too far from the aircraft's basic concept.

These were:

- the wings
  The wings were interchangeable.

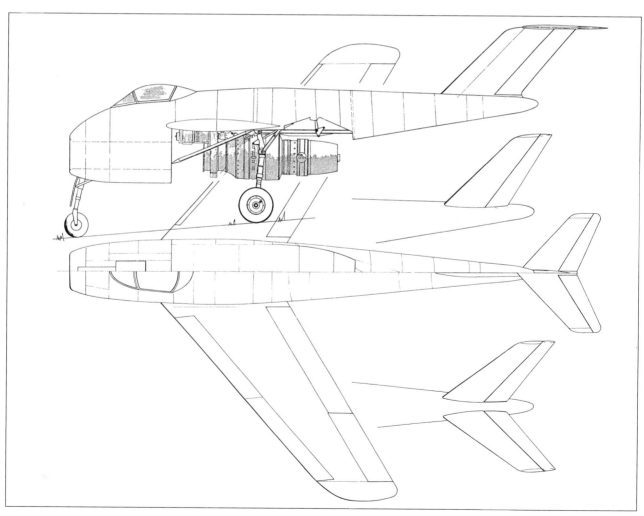

Possible modifications to the P1101 "Experimen tation Aircraft": with a T-shaped tail unit, V-shaped   tail unit, outer wing of the Me 262, wing assembly "A", Jumo 004 B, and the Heinkel He S 011.

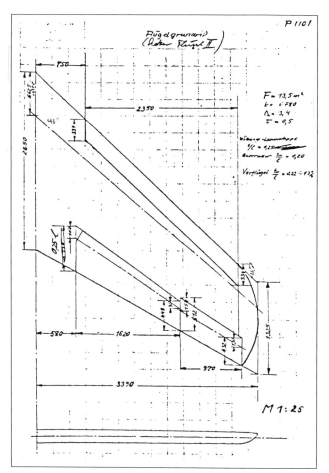

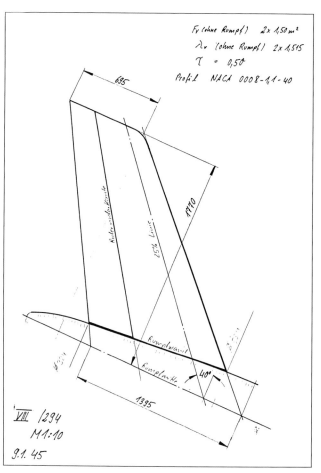

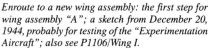

Enroute to a new wing assembly: the first step for wing assembly "A"; a sketch from December 20, 1944, probably for testing of the "Experimentation Aircraft"; also see P1106/Wing I.

Specification table of the V-shaped tail unit for the Experimentation Aircraft.

The entire fuel capacity was found in the fuselage and the shift in the center-of-gravity could be compensated for by means of a ballast.

The mounting of the wings takes place on three attachment points in the fuselage; the wing's angle of sweep could be set in three positions when on the ground (35, 40, and 45 degrees on the t/4 line). Aside from the analysis of the optimal angle of sweep, the wing was to be tested with new profiles. The wing assembly designated as "Wing Assembly A" was earmarked for the Me P1106 and Me P1110.

• the fuselage

The fuselage offered sufficient space for the installation of the most varied types of equipment, for example: electronics, weapons mock-ups, etc., and for all the fuel.

The basic design readily permitted other possibilities of the aircraft's floor plan (as, for example, the Bell engineers did with the X-5 by shifting the fuel tanks to the rear for sufficient space and for installation of an electro-mechanical setting mechanism which allowed the sweep setting of the wings to be changed

during flight)

The freedom in engine installation was obvious: the smooth fuselage allowed for the problem-free installation of either a Jumo 004B or an HeS 011; additionally, after the war, Robert J. Woods, Technical Director of Bell Aircraft Works, intended to test a whole series of American aircraft engines with the captured prototype.

• the tail unit

From the beginning, testing of the most varied tail unit shapes was planned: large and small vertical tails, horizontal tail with dihedral angle up to a pure V-shaped tail unit and a tail unit with a raised horizontal tail unit (T-shaped tail unit).

• landing gear

The landing gear and its space within the fuselage were laid out in such a way, that relatively minor modifications in installation of a strengthened or enlarged undercarriage was possible.

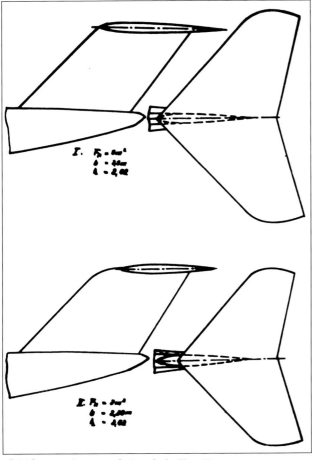

Two designs of the tail unit assembly with "high-set horizontal stabilizer." Evaluation at the AVA in Göttingen.

British swept-wing research aircraft, the Short SB-5 with the wing sweep able to be set while the aircraft is on the ground: the tail unit can be changed into a T-shaped tail unit without great expense.

The characteristics described here were an important part of the reason that, following the war, the concept of the P1101 was observed and practically unchanged by the design team of Bell Aircraft Production, which took the design and turned it into the X-5.

The English Short brothers undertook a similar design with their Short S.B.5 design.

This aircraft, constructed under the direction of David Keith-Lucas on behalf of the British government, displays clear parallels to the P1101 concept. One could change the December 2, 1952 wing sweep for the maiden flight of the "experimental aircraft" with additional equipment while on the ground for wing sweep angles of 50, 60 and 69 degrees, and the horizontal stabilizer could be placed above at the rudder (T-shaped) or deep in the fuselage. In the beginning, the British were concerned with thorough research of the problems surrounding low-speed flight, which crop up in connection with a highly swept wing.

This data then led to the English Electric P1 A and this was the test bed for Mach 2 fighter aircraft, BAC "Lightning", whose use in the Royal Air Force only came to an end in the 1980s.

As 1944 turned into 1945, production of the parts for the prototype, and the experimental construction began with the installation of the first structural components. Various parts could be taken directly from the Me 262 and the Bf 109, or be somewhat altered, (for example: landing gear, wings, controls, and other equipment), however the largest part are sheet metal and bent metal parts that were produced in the Augsburg and Oberammergau workshops.

Vehicles of the Todt Organization (OT) took over transportation of structural components and larger parts. Final assembly began in the former mule stables of the mountain infantrymen, in hangar 615, under the direction of Moritz Asam.

Up until 19/20 December 1944, trial construction activity took place in this hall with retool work on the Me 262 W-No. 170074 for the "Heimatsch͵tzer II" (Home Protector II.).

After completion, the P1101-V1 was to be brought to Lechfeld Depot near Augsburg with a special-purpose vehicle of the Organization Todt, in order assume the trial program.

Moritz Asam led a team of selected specialists with whom he wanted to construct the aircraft in an up to this point unaccustomed surface finish from Messerschmitt. It received high praises from Professor Messerschmitt, since he stood on the forefront of making the impossible possible. After the war, Asam became a conversation piece due to his decisive contribution to the design and construction of the NASA special transport aircraft Aero Spacelines "Mini Guppy" and "Super Guppy."

Those responsible used an extremely time-saving, yet risk-filled method with the 1101: production ran almost parallel to the statistical calculation, and this took place practically simultaneously with the detail construction. Except for a few small areas, delivery of the design documentation was totally completed by the end of January/beginning of February, and the 15th of March, 1945, was named the ready-for-flight deadline. Adherence to this deadline would undoubtedly be possible for the Messerschmitt Fabrik; that the aircraft would not take up its testing program until the end of the year depended mainly on the impossibility of obtaining a flight-ready engine.

Problems in the construction and modifications during construction were nothing unusual, just as they aren't today; there were also difficulties which Messerschmitt and his staff had to combat in the growing hopelessness of their situation, both economically and militarily. Allied troops were already moving on German soil. The civilian population, and therefore the Messerschmitt staff along with their relatives, and facilities, which, due to decentralization of the industry of the enormously important transportation system and most of the above ground plants and research buildings, lay in the path of the American and British hail of shells and bombs fired practically around the clock.

The Jumo 004B engine, which was badly needed for the V-1, was evidently no longer available, though both of the other required engines at the RLM for the V2 and V3 would have been obtainable. In order to undertake the least amount of component testing and adaptive work, in Oberammergau they attempted to reconstruct an engine that had been damaged and was not at all flight ready from the Me 262 Test and Operating Airfield at Lechfeld. This was apparently successful, for the Americans later registered a damaged Jumo 004 in Building 607 (mock-up).

A branch in Leonberg near Stuttgart manufactured the wings for the test aircraft. The manufacturing buildings, exclusively used for construction of the Me 262 wing assemblies, were located in a two-story constructed tunnel of the Reich autobahn and encompassed, along with above ground hangars, an area of approximately 15,000 m2. Despite the immense burden that the Me 262's mass production brought upon itself, success in carrying out construction of both sets of P1101 wing assemblies was quite easy.

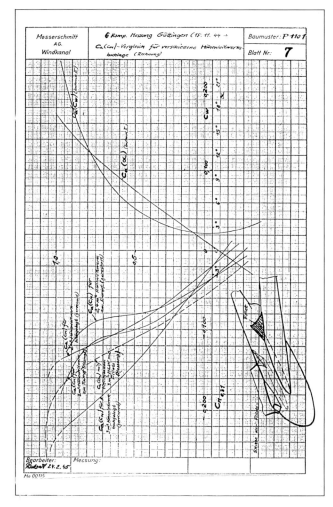

Results of the AVA measurements for various horizontal stabilizers proposed from "Fillets" between the wing's trailing edge and the fuselage. (24 February 1945).

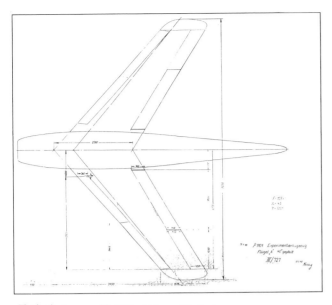

The final wing assembly "A", which was "the" wing assembly of the first successful generation of jet-fighter aircraft.

There were difficulties in transportation, and therefore the second wing assembly with slots with an increase in chord length from 13% to 20% didn't arrive in Oberammergau until the middle of February via rail.

In the so-called prototype conferences, where Professor Messerschmitt often took part, those in charge diagnosed the present state of affairs and determined further courses of action. These conferences occurred during January and February. In the official record of the "Third Prototype Conference" from the 21st of February, 1945, the following are references regarding the aircraft presently under construction:

* After some new analysis it's clear that the vertical tail to be used was too small at 0.97 m2. Thus, in the V-1, the vertical tail increased to 1.54 m2 was installed. Both of the already installed vertical tails were in Augsburg for stability and fracture testing; (a fate which saw many components from the V-1 used by Bell in the USA following the war);
* The horizontal tail was delivered according to documents.

The work and research on the "vertical tail with high-finned horizontal tail" was to draw to an end in the Project Bureau by the end of February; directly related to ending this point works was the beginning works for the V-shaped tail assembly.

* Professor Messerschmitt, who was involved with

each newly developing test bed down to each and every detail, was in agreement with construction for the new wing assembly "A" of the P1101 and P1110 wing series, as well as with testing these wings on the P1101 V1.

Meanwhile, they were already busy with the testing program for the "Experimentation Aircraft" and intended to test fly this aircraft with a wing sweep of 35 degrees; subsequently, with additional components, the wing sweep was to be increased to 45 degrees. A direct measurement of the resultant reduction in resistance as a consequence of the increased wing sweep during trans-sonic flight would have been possible ; simultaneously, flight testing would have been able to verify the effectiveness of the different measures for improving the flight-characteristics of low-speed flight.

All of these intentions and tests were carried out after the war by the Americans, the British, the French, the Soviets, and the Swedes with a multitude of "swept-wing research aircraft." A result of this powerful work is definitely the aerodynamic construction of modern jet passenger aircraft, whose flight speed is just under the speed of sound.

After the conference with the DVL in Berlin-Adlershof from 12-15 January 1945, where the Messerschmitt Firm represented the P1106 and P1110 proposals, final determination of the comparative method of calculation took place and the Technical Bureau presented the "Special Commissioned Day Fighter" and the technical and tactical requirements. Following this, it became necessary in Oberammergau to rethink the development work, and see where it all stood up to this point.

To be sure, the P1106 was not the big hit, and furthermore, in terms of performance, it offered absolutely no advantages when compared to the P1101. Therefore, Professor Messerschmitt decided to rework the P1101 mass-produced version from December 1944, to further develop the much talked about P1110, and to carry on development of a new design for a tailless fighter aircraft, the P1111 design.

Now the 1101 design received the more powerful weaponry, increased fuel capacity and an improved outline. (For details, see the section: Mass Production of the P1101).

The three designs submitted and spoken about at the meeting from 27 and 28 February received the best assessment, along with the Junker's EF-128 proposal.

The aircraft designated as the Tank Ta 183, often referred to in post-war literature as the victor, did not receive any attention at this point in time due to its insufficient maximum speed; for more, see the chapter "Evaluation of the Designs."

A final decision on a future aircraft series did not occur, and it certainly did not occur in the few weeks leading up to the end of the war. The RLM only awarded the manufacturing contract for the prototype aircraft, and such was also the reasoning process from this conference: "The test aircraft, presently in construction, is to be completed as quickly as possible and tested the same; a similar process is valid for other test aircraft presently under construction."

However: the end of the war was waiting just around the corner, and living conditions worsened noticeably, and the working conditions adopted a chaotic form. Organized work was long since unthinkable, material was unobtainable, air raid alarms often startled staff members, and they were forced, packed with their most important documents, into underground parts of the plant, or to seek protection in the barely covered terrain outside the research institute. Characteristically, the staff did not flee into nearby forests, for the aggressors either knew or suspected, that the Messerschmitt staff members still had painful memories from Augsburg.

Lucky for the staff members, the Upper Bavarian Research Institute did not experience a single air raid during the entire war. Being tucked away and the "disguise" was evidently carried out perfectly. Was it real? There are legitimate doubts.

Perhaps the French news service received exact information either via the Parisian Messerschmitt Bureau or via the large number of French active in the work about the work at the Upper Bavarian Research Institute; information which was suppressed from their own allies, and which, a few days before the end of the war, would prove to be invaluable for France.

This acceptance of such is bolstered as a leading staff member from the Parisian office, following the war called himself a "resistance fighter from the very beginning", and because some of the French workers had organized themselves into resistance and sabotage groups and held a connection with the resistance movement in France.

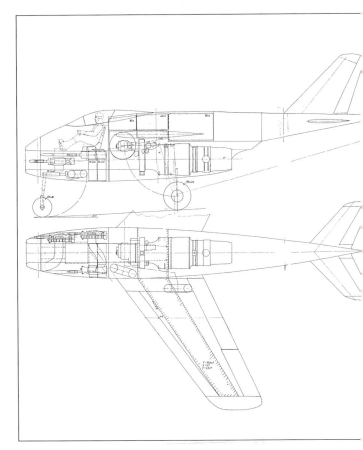

Enroute to the final P1101 production series proposal; Alternative to the current landing gear kinematics: P1101 with the retractable undercarriage to the front (February 20, 1945).

In March 1945, the required operationally ready Junker's jet engine was still not to be found in Oberammergau, and it still wouldn't arrive by the end of the war.

After the DVL had criticized the P1101 production series proposal's upper tail unit over the wing, the Project Bureau manufactured the design configuration for the "tail unit with high horizontal stabilizer" for the experimentation aircraft as a basis for the detail construction; otherwise the staff members were totally engaged for Professor Messerschmitt and Engineer Voigt.

Somewhat abbreviated, the main work dealt with:
- the gradual improvement of the production series (inclusion of desires and complaints of the forces);
- the further development and conception of the Me 262 variants (high speed, night-fighter, etc.);

- the development of the single-engine jet fighter (P1101, P1110, P1111);
- and last but not least, the revolutionary P1107/ P1108 jet bomber program, which came to its demise after the war in the British specification B35/ 46 (where the Arro "Vulcan" emerged), and in the commercial aircraft D.M.106 Comet. Finally, de Havilland's Chief Engineer, R.E. Bishop, was not found in June 1945 in Oberammergau as a mere vacationer!

But highly cultivated piston engine fighter delivered the last classic air battles in the skies over the defeated Reich. A revolution in aviation technology occurred with the end of the war: a revolution which had already thrown its shadows before the war, but which, in its radicalness, is totally without example in the history of the aviation.

In April 1945, the 1101-V1 was about 80% finished. A damaged Jumo engine was even available for modification work, however a disguise could no longer be built. In the meantime, the first flight was set for June 1945, though for obvious reasons, the project was too expensive.

Nevertheless, at least parts of the V1 flew, though unintentionally: While copying the cockpit (testing the sealing with light overpressure), the cockpit canopy of the prototype flew with one loud bang on the ceiling of hangar 615.

On 27 April, one of the few Messerschmitt workers left in Oberammergau, along with a few French assistants, pulled an almost finished P1101 from the assembly hall.

The workers created the airplane in the underground section of the plant in order to prepare it there for explosions. But it never did occur.

Technical description
The P1101-V1 Experimentation Aircraft: Construction of the first test bed was accomplished by the Oberammergau Bureau under the direction of Walter Rethel between December 1944 and January 1945, ergo, within 6-8 weeks.

Due to time constraints, the first prototype served simultaneously as a "dummy" for diverse modification work. These facts, hidden by the then leader of

P1101-V1, condition after the invasion of the U.S. armed forces; Model construction: G. Sengfelder (above).

Middle Picture:
Model comparison Me 262 - P1101 (on the left).

Model comparison He 162 - P1101 (below) (Model construction: Sengfelder).

dummy construction, were considered conjectures and mistakes by some chroniclers after the war.

Involved directly in the origin of the P1101-V1 were:

Design: Messerschmitt, Voigt, Hornung
Construction director: Keidel

Construction:
Control:	Tethel
Fuselage:	Barmeyer
Wings:	Weissflog
Undercarriage:	Schäffler
Tail unit:	Bosch

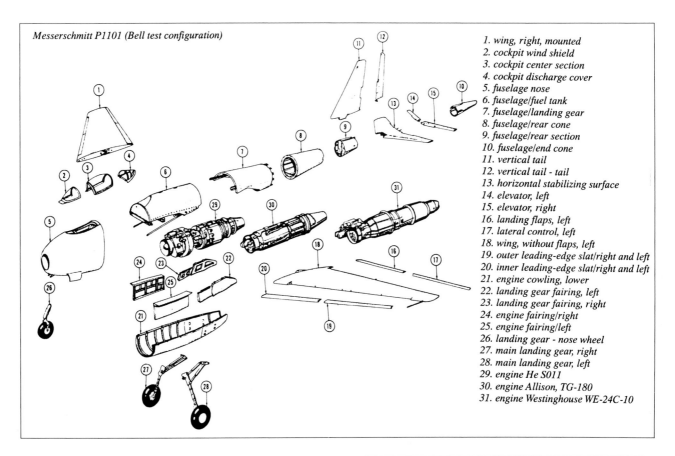

Messerschmitt P1101 (Bell test configuration)

1. wing, right, mounted
2. cockpit wind shield
3. cockpit center section
4. cockpit discharge cover
5. fuselage nose
6. fuselage/fuel tank
7. fuselage/landing gear
8. fuselage/rear cone
9. fuselage/rear section
10. fuselage/end cone
11. vertical tail
12. vertical tail - tail
13. horizontal stabilizing surface
14. elevator, left
15. elevator, right
16. landing flaps, left
17. lateral control, left
18. wing, without flaps, left
19. outer leading-edge slat/right and left
20. inner leading-edge slat/right and left
21. engine cowling, lower
22. landing gear fairing, left
23. landing gear fairing, right
24. engine fairing/right
25. engine fairing/left
26. landing gear - nose wheel
27. main landing gear, right
28. main landing gear, left
29. engine He S011
30. engine Allison, TG-180
31. engine Westinghouse WE-24C-10

P1101-V1 with its main components disassembled: Figure: Aerofax/Jay Miller (above).

Fuselage nose of the V1; pictured after casting in Oberammergau, Hangar 615. The damage caused by the war is clearly visible (right).

The typical Messerschmitt side hinged cockpit canopy (below).

Engine installation and fuel tank
  construction: Beese
Hydraulic and electric equipment: Harsh
Ground facility equipment: Zimmerman
Communications package: Menhardt
Firmness/statics: Ebner
Test construction: Asam

When building the P1101-V1, special emphasis was placed on an extremely clean construction and a flawless surface finish (rivet heads and butt joints do not protrude). This was and still is imperative for attainment of the highest possible speeds, and additionally, production had, in this regard, something to compensate for; it was the construction of the Me 262 that was exposed to sharp criticism.

According to the Prototype Director, Walter Keidel, the aircraft would have been flight ready within 2-3 weeks at the time of the invasion.

Fuselage

The fuselage is constructed out of duralumin. The fuselage fairing is lined for the Jumo 004-B on the first test aircraft; retooling the Heinkel HeS 011 engine is possible without difficulty. The basic dimensions and body composition of the V1 fuselage are comparable to the later production series.

The fuselage forebody contains the air transfer tube, the nose wheel, which is retractable to the rear, and the space for the pilot. The loss of engine thrust through the air transfer tube was analyzed with an intake scoop similar to one of these tubes on the left engine of an Me 262.

The space for the nose wheel is already designed for the larger wheel from the production series; and there is also room enough for the possible installation of weaponry and for equipping it with a pressurized cockpit. With the cockpit canopy, the concern is with a full view dome, whose construction was designated somewhat popularly at Messerschmitt as a "racing cockpit."

The windscreens of the cockpit are kept fog-free through circulation of the warm air drawn from the engine. A dome of similar construction was already tested with an engine retooled identically to the Me 262.

The entire fuel capacity is housed (1000 liters) in the middle section of the fuselage; though not in protected tanks, but rather in four tightly riveted fuselage sections.

The engine lies underneath the fuel tank area. Through a reduction of the entire engine housing, the engine is easily accessible without specialized equipment and is thereby easily maintained; the engine and fuel are separated by a fire-proof bulkhead.

The space foreseen for the main undercarriage is also already set for the larger wheels of the production series.

The cone-shaped fuselage end houses the radio equipment, the oxygen system, the directional control, and the master compass. Because of the thermal radiation from the

engine, the sheathing on the underside of the fuselage is fitted with steel plating. And since the master compass cannot tolerate steel, it's slid to the extreme end of the fuselage and is covered by a hood of non-steel material.

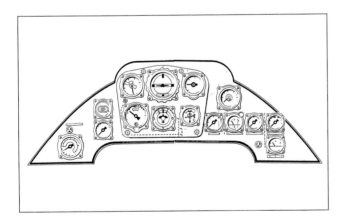

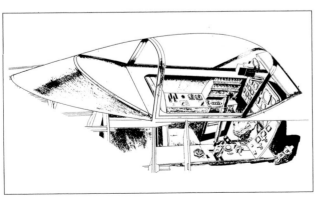

Instrumental panel of the P1101-V1 (left above): The setup of the left and right instrument boards correspond right up to the "switchbox for launching" of the Me 262 A-1.

(above)

Tail unit with the aft fuselage of the P1101 (above).

(lower left)

Pilot's cockpit of the Me 262 A-1, identical to the cockpit of the P1101-V1. (below, left)

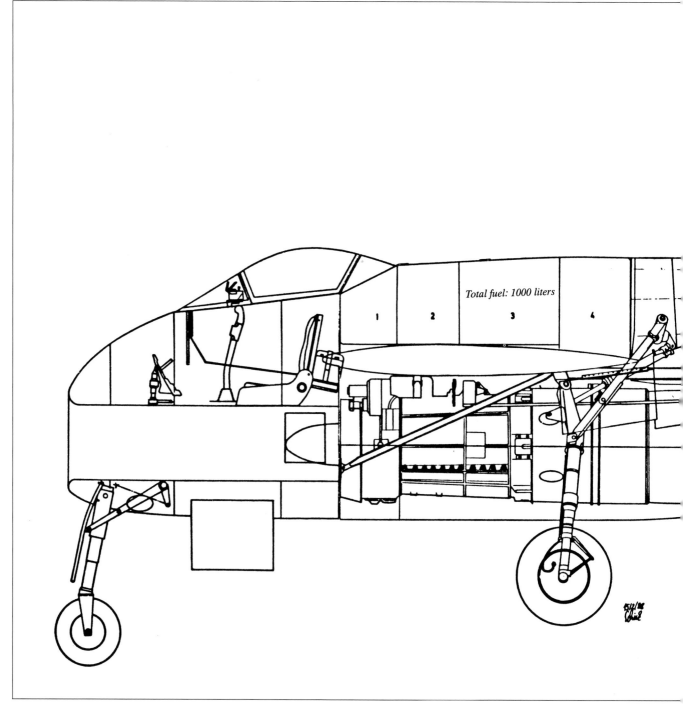

Total fuel: 1000 liters

A fuselage section form of the tail unit serves for maintaining the communications equipment, the direction control, and the oxygen system.

The wing assembly

The first wing assembly set of the experimentation aircraft is somewhat similar, in the outer wing, to that of the Me 262 in the area from rib 7 (engine) up to rib 21 (final rib). The wing tip edge is adjusted to a 40 degree wing sweep. The profiling of the wing assembly, the aileron section, and the leading edge flaps (slots) correspond to those of the Me 262. The designers had hoped to counter the difficulties of the swept wing during low-speed flight by employing leading edge slats over almost the entire wing span; a solution which called forth some uncertainties, however:

66

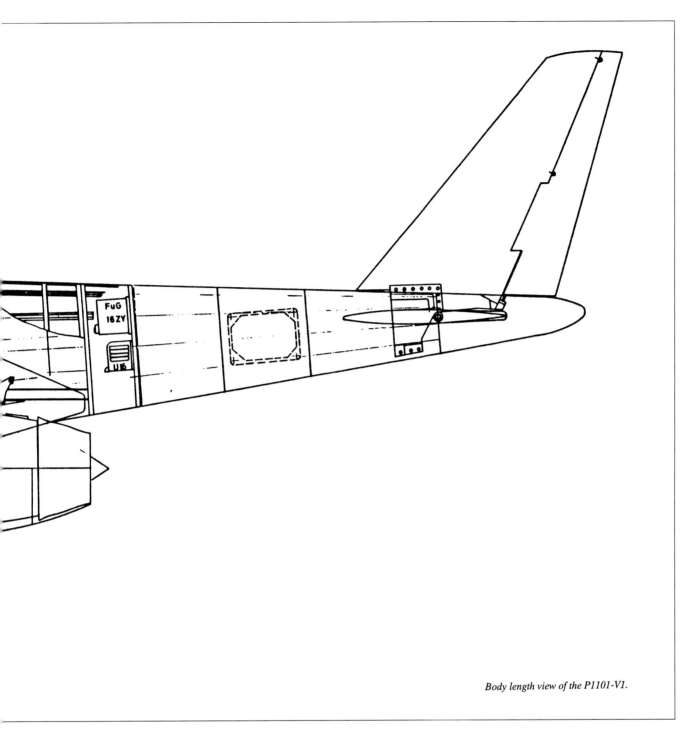

Body length view of the P1101-V1.

With the second wing assembly test set, delivered in February, 1945, to the workshops in Leonberg, the slot (leading edge flaps) had been enlarged from 13% to 20% of the wing chord. A solution, which, despite the concerns of the RLM ("in high-speed flight, the leading edge flap also maintains a considerable drag increment, since the wing profile is interrupted at the most sensitive part of the flow at the upper front profile contour"), was aerodynamically forward-looking, as several post-war swept wings are constructed with such lift-increasing devices have proven (leading edge flaps and slat wings). The design philosophy taken from the Soviets by Focke-Wulf using the low wing capacity without relatively complicated lift-increasing devices was realized by the first generations of jet fighters with plenty of steel in the boundary layer fences on swept wings.

But the newer constructions were given up in favor of an engine that was more modifiable and had similar components.

The landing flaps of the P1101 are employed as plain flaps. At the expense of the landing flap, the aileron can be extended if the construction caused too tight of a design.

The wing is attached to the fuselage at three points.

One particular feature of the test aircraft was already mentioned: the possibility of the wing's angle

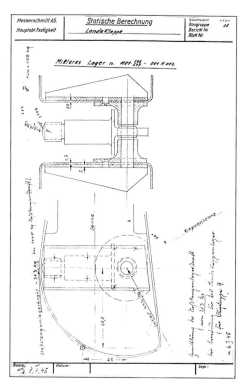

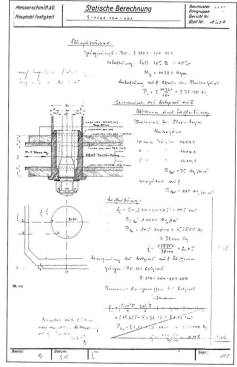

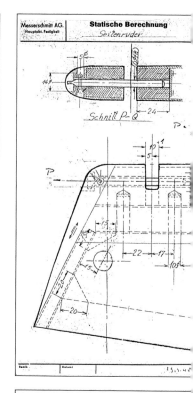

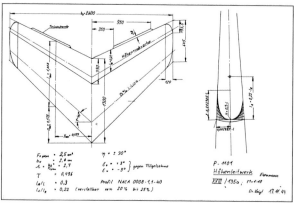

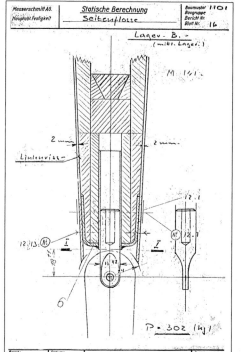

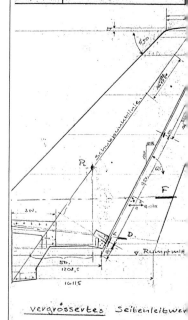

The middle joint of the landing flap (above, left).

Wing - fuselage connection, main spar connection. An example from the statistical computation (above, middle).

Specifications of the rudder (below, right), rudder connection, tail unit in its wooden form (right, middle), rudder joint (above, right).

Diagram overview of the horizontal tail unit; from the project delivery (above).

of sweep to be set in three different positions while the aircraft is on the ground (35, 40, and 45 degrees). The fuselage side/front spar union must be swapped when changing the wing sweep to one of the afore-mentioned angles.

The connection at the main spar is accomplished by means of a heat-treated, vertically set steel bolt into the fuselage side, with the 2731 frame-connected cover plate; the wing is angled at this point.

In static composition, the P1101 wing is identical to the Me 262 wing: Along with the rear auxiliary spar and the mounted light-metal ribs with intervals of 270 mm, a full wall beam for a main spar with a double-T profile forms the frame for the stressed skin, riveted into the panel.

The tail unit

The tail unit is a normal unit built in a cruciform configuration. The vertical tail as well as horizontal tail units are constructed out of wood and have a sweep of 40 degrees on the t/4-line.

The so-called "large vertical tail" is installed on the prototype. The rudder is supported to the fin at three points. The deflection angle of the rudder is limited to ±20 degrees at full pedal stroke, while the elevator angle can withstand ±30 degrees.

Steering

The mechanical steering was largely that from the Me 262.

The undercarriage

The nose wheel, which was unable to be used for braking, retracted to the rear, and was sized 465 mm x 165 mm, had an enclosed, hydraulic flutter damper due to its turning ability while on the ground. The hydraulic cylinder for the retraction mechanism works via a radius strut.

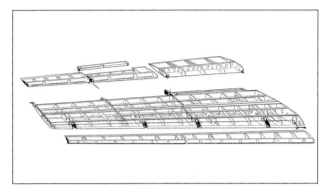

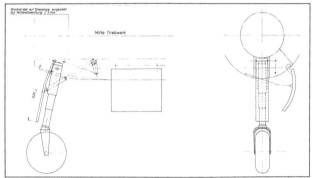

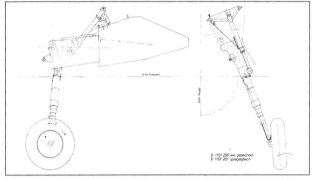

Overview of the wing assembly P1101-V1: 1st wing set with slots at 13%

Nose wheel landing gear of the 1101-V1

Main landing gear of the 1101-V1

(from top to bottom)

A functional model of the undercarriage while extended (left).

The main landing gear originates, to a large degree, from the Bf 109K.

The 660 mm x 190 mm wheels are capable of being braked. Storage of the landing gear leg and hydraulic retraction mechanism were newly constructed with the Bf 109K in mind.

The engine

With the first installed engine for testing, the concern is with the mass-produced, already tested and operational (in the Me 262, Ar 234) Junkers Jumo 109-004 B. As soon as the Heinkel HeS 109-111 A, with more thrust, was available, an exchange was planned for this engine.

Equipment

No military equipment, such as armor-plating or weaponry, are planned for the experimental aircraft. The aircraft is equipped with the fighter control of heading of the Me 262. For radio equipment, the remote controllable VHF radio FuG from Lorenz is planned.

Functional model of the retracted undercarriage (above).

Main undercarriage with a mock-up of the He S 001 engine. (on the right)

P1101-V1 wing assembly-fuselage mid-section and main undercarriage. In this picture, which emerged from the Bell Aircraft Works in Buffalo, N.Y., in 1948, the damage which occurred in the invasion and then during transport is clearly visible. (below)

The instrument panel is lined up as follows:
Instrument flight panel I with:
Air speed indicator
Turn and bank indicator with artificial horizon
Rate and climb indicator
Altimeter
Pilot's auxiliary compass
Automatic frequency tuning 2 indicator
with monitoring instruments for the engine:
RPM indicator
Gas pressure meter
Injector pressure gauge (fuel)
Lubricant pressure indicator
Fuel capacity indicator
For the on-board power supply there is a 3000 watt generator.

Technical data

Primary Function:
Swept-wing research aircraft, possibly a prototype for an aircraft series
Crew:
1 pilot in a pressurized cockpit capable of withstanding Mach flight
Length:
8980 mm
Height:
Approximately 3500 mm
Wing assembly:
(Derived from the Me 262 "Wing Assembly A", see P1101 series)
Wing assembly surface:
13.6 m^2
Wing assembly profile:
Root: National Advisory Committee for Aeronautics 00011.41 -1.1 -40
Wing nose point: NACA 00009 -1.1 -40
Aspect ratio b2/F:
= 4.78
Sweep: (adjustable on the ground)
(t/4-line): 35/40/45 degrees
Wing span:
8060 mm (with sweep of 40 degrees)
Dihedral angle:
Normal position: adjustable from 0-3 to + 2 degrees
Angle to fuselage axis:
0 degree wing assembly without aerodynamic decalage
Undercarriage:
Nose wheel: 465 x 165 mm
Main wheel: 660 x 190 mm
Wheel base: 2124 mm
Camber 4 degrees 20' inclined toward fuselage
Tail unit:
Vertical tail:
Total surface: 1.54 m2
Surface, directional control: 0.40 m^2
Profile: NACA 0008 - 1.1 - 40
Horizontal tail:
Wing span: 2650 mm
Total surface: 2 x 1.20 m2 = 2.40 m2

Rudder surface: 2 x 0.30 m2
Profile: NACA 0008 - 1.1 - 40
Engine:
1 x Junkers Jumo 109 - 004 B maximum thrust:
B-1 kN (910 kp) B-4 9.81 kN (1000 kp)
Weight:
Weight of components during construction:

Fuselage nose	89.0 kg
Cockpit1	5.4 kg
Middle fuselage section	60.0 kg
Tank 1 + 4	160.0 kg
Landing gear beam	36.0 kg
Mount	5.0 kg
Rear fuselage section	40.1 kg
Seat	10.0 kg
Fuselage (approximately)	416.0 kg
Horizontal tai	172.3 kg
Vertical tail (large)	40.8 kg
Tail unit (approximately)	113.0 kg
Main landing gear	224.0 kg
Nose wheel	41.8 kg
Undercarriage (approximately)	266.0 kg
Wing assembly with Components	350.0 kg
Hinged back plate	60.0 kg
Wing assembly	410.0 kg
Control in fuselage(approx)	35.0 kg
Equipment with instrumentation	6.8 kg
Instrument panel, left	2.4 kg
Instrument panel, right	5.2 kg
Oxygen system	5.0 kg
Wiring and fittings	1.2 kg
Oil pressure system/nose wheel	14.2 kg
Breaking device in fuselage nose	0.7 kg
Electrical systems in fuselage nose	13.2 kg
3000 watt generator	15.5 kg
Accumulator 24 V/20 ampere hours	24.0 kg
Hydraulics, brakes for main landing gear	13.3 kg
Service equipment (approx)	103.0 kg
Engine cowling	50.0 kg
Engine controls	3.1 kg
Fuel system inside engine	3.0 kg
Jumo 004 B engine	750.0 kg
Engine	806.1 kg
FuG 16 ZY (radio)	35.0 kg
Equipment	35.0 kg
Empty weight	2184.0 kg
Pilot	100.0 kg
Fuel (1000 liters)	830.0 kg
Ballast	91.0 kg
Removable load	1021.0 kg
Take-off weight:	3205.0 kg

Flight performance: (calculated)
Maximum speed: 860 km/h at altitude of 7 km
Diving speed: 1100 km/h
Climbing speed: 12 m/sec at 0 altitude
Landing speed: 170 km/h
Landing distance: 900 m
Operating ceiling: 10,000 m
Wing load (take-off): 236 kg/m2

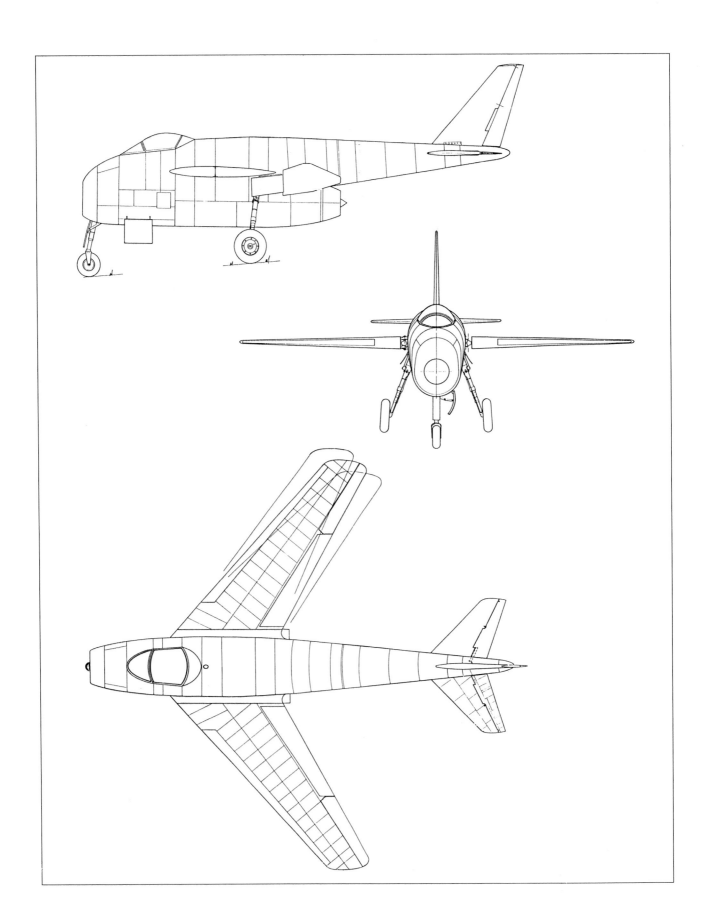

The actual capabilities of the P1101-V1 might be somewhat higher, as the following comparison illustrates with the Soviet counterpart to the P1101, the Lavotschkin La-160.

In the fall of 1946, ground testing was performed on an aircraft undertaken in Russia which greatly differed in comparison to all designs built up to that point in Russia. The one-seater aircraft, mimicking German construction, is powered by an afterburning engine and had, as a spectacular feature, a thin, swept wing. And a built-in ejection seat represented a further, unheard-of feature for Soviet aviation.

With this "Experimentation Aircraft", designated as Lavotschkin La-160 "Strelka" (Arrow), the Soviets could carry out an extensive testing program after the first successful flight on June 24, 1947. With several already built aircraft, scientists and engineers had achieved valuable knowledge, which would later be introduced into the construction of future mass-produced aircraft.

The La-160, built with the aid of German documents, did not reach the aerodynamic refinement of the P1101. Still, it appears doubtful whether Messerschmitt documents were found useful at all during construction of the Russian swept-wing aircraft. Therefore, a comparison of both designs is very informative.

P1101 Production Series
(Status at the end of February 1945/RLM documents).

Fundamentally, the planned 1101 production series had the same construction as the experimental aircraft. Due to material shortages, however, the abundance of wood saw it, to a large extent, as the only intended construction material.

Naturally, the equipment for an operational aircraft is essentially more important than for a test aircraft. At this point in time, the beginning of 1945, the equipment layout from January 2, 1944, can be considered, in part, as having been replaced.

As for aircraft weaponry, the movement was clearly toward long-range aerial rockets in order to interdict the aggressor through strong defensive weapons of the bombers, a tendency which was still seen for the years following the war and right up to the present day. The HS 298 radio-controlled, guided missiles from the equipment layout of November 1944 was already given up in favor of the wire-controlled, X-4 air-to-air Ruhrstahl rocket. With unguided rockets, more and more the construction type with rear stabilizers prevailed over those with high-caliber, launchable projectiles with spin stabilization.

P1101 in comparison:	Lovatschkin La-160 "Strelka" ("Arrow")	Messerschmitt P1101-V1
Application	Swept-wing research aircraft	Swept-wing research aircraft
Engine	1 x RD-10F with with 1100 kp thrust (modified Jumo 004-A engine)	1 x Jumo 004 B 890 kp thrust
Length	10.06 meters	8.98 meters
Wing span	8.95 meters	8.06 meters
Wing surface	15.9 meters	13.6 meters
Aspect ratio	5.0	4.78
Dihedral angle	- 2 degrees	- 3 degrees up to + 2 degrees
Profile	TSAGI- maximum speed	NACA series
Relative wing thickness	9.5%	average 10%
Sweep	35 degrees	35 degrees - 45 degrees
Empty weight	2738 kg	2184 kg
Take-off weight	4060 kg	3205 kg
Fuel	approx. 1500 liters	1000 liters
Max. wing load	255 kg/m2	236 kg/m^2
Thrust load	3.7 kg/kp	3.6 kg/kp
Max. speed	900 km/h at ground level 1060 km/h at altitude of 5,700 m	860 km/h at 7,000 m altitude
Operating ceiling	11,000 m	0,000 m
Weaponry	2 x 37 mm Mk N-37	Ballast
Aerodynamic aids	2 boundary layer fences on each wing half	Leading edge flaps over almost the entire wing span
Maiden flight	June 24, 1947	planned for June 1945

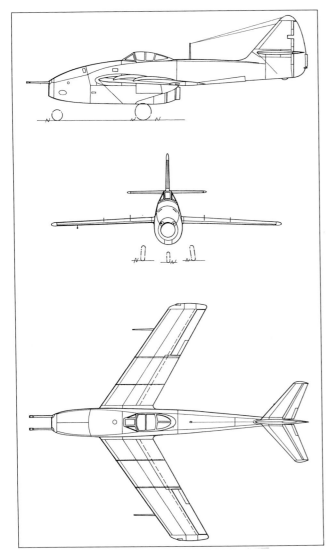

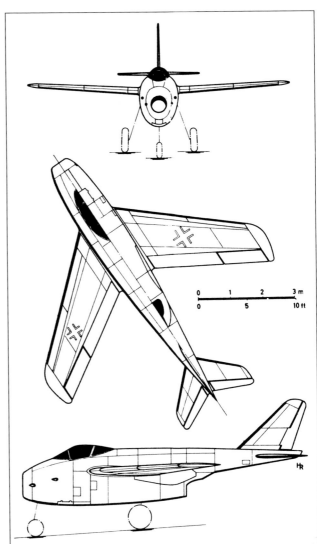

The Soviet 1101: Lavotschkin La-160 (above)

Full-view drawing of the P1101 series (right)

The greatest progress in the area of electronic warfare came from the machines of the adversaries. Therefore, in early 1945, it was planned to use electronic equipment that would compare to the spring of 1944 but would be modified, improved, and sometimes also simplified.

Naturally, the weapons technology and electronic equipment of an aircraft in production succumbed to almost constant modification. The condition described in the appendix is the probable technological status of equipment for the fighter-interceptor of the Luftwaffe (1) from 1946/1947 and it is similar in many areas to many post-war fighters.

(1) Note: Of course, this is in view of the comprehensive circumstances of the events of the war and under the assumption that the described weapons and equipment would be proven operationally suitable, which, in part, never underwent serious testing.

Fuselage

Construction of the fuselage as all-metal would be similar to the construction of the prototype's fuselage. The shape is somewhat improved, and it was planned from the beginning to install: 4 Mk 108; a Heinkel HeS 011 engine; and three armored tanks for taking in the largest possible amounts of fuel.

The crew compartment is a pressurized cabin. The pilot is protected through armor plating. (From the front against 12.7 mm rounds and from the rear against 20 mm rounds).

Wing assembly

The wing assembly is a two-piece wooden-surface with a sweep of 40 degrees. This sweep is unchangeable. The automatic leading edge flaps mounted to the wing nose have different percentage wing chord settings over the wing span. Thereby, the cross section of the trailing edge flap remains almost constant.

The landing flaps are employed as trailing edge flaps.

In the wing assembly there are unprotected areas for fuel.

The aerodynamic construction is, in a large part, similar to the already described "Wing Assembly A", which features an increase in the relative wing thickness from the root (8%) to the wing tip (12%). Coupled with the leading edge flaps, the Messerschmitt engineers had hoped to reach acceptable low-speed flight characteristics through these measures; in addition, this construction type had sturdiness and technical construction advantages.

The described wing no longer has similarities with the Me 262.

The foreseen wooden construction would prove unfulfillable by the end of the war, despite the availability of materials: the required wood cement was no longer available as a result of the bombing raids and a substitute could hardly be used, as the debacle of the Tank Ta 154 night-fighter illustrated.

Tail unit

The vertical tail as well as the horizontal tail units are, just as the wing assembly, constructed from wood and swept to the 0.25 t-line to 40 degrees.

Undercarriage

Naturally, due to the higher take-off weight of the production series compared to that of the prototype, the undercarriage is reinforced. The principle composition is roughly the same. The nose wheel, 500 x 180 mm, is rotated approximately 90 degrees when retracted.

The main undercarriage, with a wheel sized 740 x 210 mm, has a retractable, pneumohydraulic shock absorbing strut.

Engine

For the engine of the P1101, the question at the end of the war in all studies, proposals, and in all designs by German aircraft builders was with the Heinkel-Hirth HeS 109-011 A-0 engine (HeS 011 for short).

Technical data of the P1101 production series aircraft
(Status as of February 22, 1945)

Role:
Fighter, interceptor, and fighter-bomber
Crew:
1 pilot in a pressurized cockpit
Engine:
1 x Heinkel-Hirth HeS 109-011 A-0
Thrust 1300 kp (12.8 kN)
Full-throttle thrust 1020 kp at ground level
100%   480 kp at altitude of 10 km
Length:
9175 mm
Height:
3710 mm
Wing assembly:
Wing assembly "A" is similar aerodynamically to the wing assembly of the P1106 and P1110; slight differences in geometric data only
Wing assembly profile:
Root NACA 0008-1, 2-40
Wing tip NACA 0012-1, 0-40
Dihedral angle: 0 degrees
Tapering: 0.524
Angle to fuselage axis: 0 degrees
Wing assembly without aerodynamic decalage

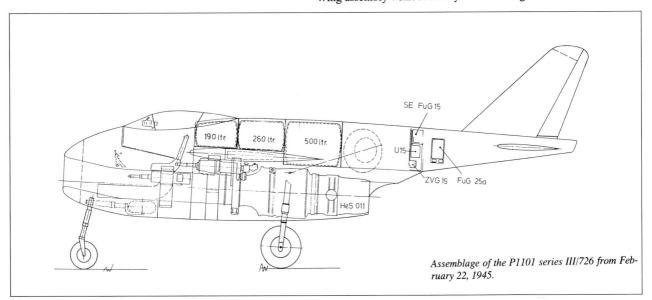

Assemblage of the P1101 series III/726 from February 22, 1945.

Undercarriage:
Nose wheel: 500 x 180 mm
Main wheel: 740 x 210 mm
Wheel base: 2200 mm
total allowable static load: 6000 kg

Vertical tail:
Surface with rudder: 1.4 m^2

Horizontal tail:
Wing span: 2580 mm
Surface with rudder: 2.45 m^2

Weights:
Data according to Engineer Hornung from June 27, 1945
Wings   450 kg
Fuselage   300 kg
Tail unit with control   128 kg
Undercarriage   230 kg
Total Aerostructure   1108 kg

Total Aerostructure   1108 kg
Engine with attachments   986 kg
Equipment   190 kg
Armament (Mk 108 with armor)   310 kg
Empty weight   2594 kg

Pilot   100 kg
Fuel   1250 kg
Munitions   120 kg
Payload   1470 kg

Normal take-off weight   4064 kg

Thereby the take-off weight, mainly because of the increased fuel volume, is approximately 200 kg more than the weight from December 1944 (3863 kg).

The P1101 leans toward the 5000 kg maximum take-off weight class, right next to the last, classic, piston motor aircraft; in comparison, the Spitfire F Mk 22 had a maximum take-off weight of 4865 kg; P-51 D Mustang a maximum take-off weight of 5490 kg; and even the Me 163 rocket fighter came to the scales with 4310 kg. With the exception, perhaps, of Soviet designed aircraft, the first post-war generation of jet-fighter aircraft had almost double the take-off weight as the Messerschmitt design.

One can easily accept, in view of the 1101 and surely the 1110 designs, that Messerschmitt fully succeeded in fulfilling the original RLM requirement for "a small turbo-jet engine, single-seater with the highest possible flight performance": and speaking of performance characteristics:

Flight performance characteristics:
(according to DVL industry calculations)
Maximum speed
at sea level   885 km/h
at 7 km altitude   985 km/h
Climbing speed
at sea level   22.2 m/sec
Climb time
to 2000 m   1.5 minutes

Taxiing out distance with take-off 700-750 without boosters
Landing speed with 1/3 fuel remainder   172 km/h
Taxiing distance during landing   570 m

The following data comes from an American International Committee of Scientific Management report:

Operating ceiling   12,000 m
Flight time at ground level   40 minutes
Flight time at 10,000 m altitude   1.8 hours
Operating range at ground level   500 km
Operating range at 10,000 m altitude   1500 km

Despite the extraordinarily short development time of approximately one-half year, the P1101 had reached a certain level that would have allowed for the acceptance of its draft and thereby immediate construction, and would ultimately make it stand out as the first jet fighter immediately following the war. Out of all of the test and production series aircraft of the early generation of jets, the Swedish SAAB J-29 was at the forefront, alongside the Sabre F-86 and the MiG-15. And like American and Soviet test aircraft, the J-29's roots lead, in many respects, back to German designs.

This first, ready-for-action, swept-wing fighter was Western Europe's successful product from masterful, well thought-out Swedish engineering work, German aerodynamics, and dependable British engine technology.

With the J-29, not only was the swept wing used for the first time on an operational aircraft, but the unwieldy fuselage, through the expanded, centrifugal engine was designed according to the area rule concept; this is one of the reasons the J-29 was hardly inferior to the slimmer F-86 in flight performance.

Interval

The apparent alternative - The 1106 project:

The further the work on the P1101 production series progressed in the autumn of 1944, the more often and more grave the problems proved to be and the more doubts began to surface.

First and foremost, the difficulties were:

- the extremely condensed weapons installation space, which only allowed for limited weapons capacity;
- the complicated housing for the main undercarriage (For this, Professor Messerschmitt proposed, in the middle of October, the P1101 with the forward-retractable undercarriage; this design was worked out later in a detailed layout);
- the unreasonable order for the principle load, which causes the fuselage weight to increase somewhat;
- and, perhaps the gravest problem — the calculated maximum speed was lower than anticipated. Furthermore, the calculation contained a few

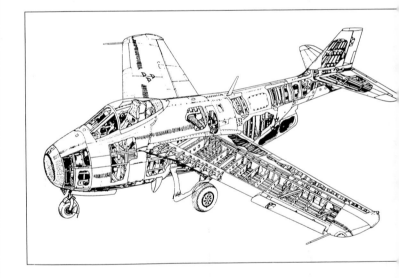

SAAB J-29 A, first production aircraft W. Number 29101. (left)

SAAB J-29 A cross-section drawing. The P1101's influence is unmistakable. (above)

unweighable factors, due to the chosen fuselage shape, which, with its high frame, declined considerably from the spindle shape with a spherical profile;
- limited by the requirements for the engine, the moment of thrust produced considerable displacement of indifference points between full-power and idle (trimming problems);

P1101 in comparison:	SAAB J-29 A	Messerschmitt P1101 series aircraft from February 2, 1945
Application	Fighter	Fighter
Crew	1 pilot in pressurized cockpit with ejection seat	1 pilot in pressurized cockpit with ejection seat being tested
Engine	Aero-engine RM2 (License D.H. Ghost) with 2270 kp static thrust	Heinkel-Hirth HeS 109-011 A-0 with 1300 kp static thrust
Length	10,227 mm	9175 mm
Height	3750 mm	3710 mm
Undercarriage wheel base	2200 mm	2200 mm
Wing span	11,000 mm	8250 mm
Wing area	24 m2	15.85 m2
Aspect ratio	5.0	4.29
Profile thickness	8.5%	8% 12%
Sweep t/4 line	25 degrees	40 degrees
Aerodynamic aids	slots in the outer wing the high sweep near fuselage	slots over almost entire wing span
Empty weight	4580 kg	2594 kg
Normal take-off weight	6880 kg	4064 kg
Maximum take-off weight	7530 kg	4700 kg
Normal wing load	286.7 kg/m2	256 kg/m2
Maximum wing load	313 kg/m2	296.5 kg/m2
Normal shear load	3.03 kg/kp	3.1 kg/kp
Fuel	1430 liters and 900 liter auxiliary tank	
Armament	4 x 20 mm Mk and 650 kg of external load	4 x 30 mm Mk and 500 kg of external load
Performance Maximum speed	1035 km/h	985 km/h
Climb time at 10 km altitude	7.3 minutes	approx. 9.5 minutes
Operating ceiling	13,700 m	approx. 14,000 m
Landing speed	220 km/h	172 km/h
Take-off distance	900 m	750 m
Landing distance	600 m	570 m
Operating range	1200 km without auxiliary fuel	maximum 1500 km

All of this and some new occurrences from the research on high-speed flight caused those responsible to propose further developments for the P1101 design.

In this case, further development meant, above all, to rid the design of the speed hampering and fuel consuming problems.

For the fuselage breakdown of the design labeled P1106, the project engineers chose a new requirement for the main structural components:

The engine's air intakes were worked on almost immediately, with the goal of keeping the intake to loss of thrust ratio to the absolute minimum; viewed roughly, the 1106 design possessed about half the intake length of the 1101.

In one experiment (for a "new project", probably the 1101), they wanted to ascertain the amount of thrust loss originating from the nose intake of an engine housed in the fuselage.

Experiment Engineer Kaiser, testing a 3-meter long intake under unfavorable conditions, determined a maximum loss of 3%. An advantageous design was determined to be a smooth valve with a round intake construction.

Experiment to measure the intake loss of a fuselage-housed engine, ascertaining the optimal intake design (November 15, 1944, above).

First design for the P1106 from December 4, 1944 (below).

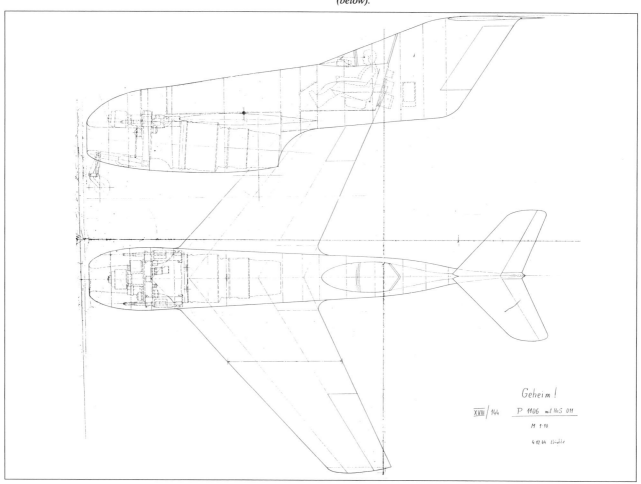

To accommodate this, the cockpit had to be slid quite far back to the leading edge of the tail fin. The jet exhaust, in turn, has the shape of a graduated, semi-circle.

From the beginning, the P1106 received a somewhat larger capacity for fuel (P1101: 1079 kg, P1106: 1200 kg). Below the fuel tank, which occupied the upper half of the fuselage between the armament and the pilot, the main beam of the wing passed right through. The armament was located, on the other hand, in the fuselage nose; the space proportioning, through the omission of the cockpit, makes this somewhat more acceptable.

With the first P1106 design (dated December 4, 1944), a forward-placed "vertical tail with high-set horizontal tail" had been planned.

The somewhat tediously formulated structure is generally known nowadays as a "T-shaped" tail unit.

With this novel idea, the full reality of the horizontal tail unit for trans-sonic flight should be possible: the horizontal tail unit would be placed up top due to the "wake zone" from the separated-flow shock of the wing and fuselage. The aim of the designers was to reduce the shock waves through elimination the local overspeed or to totally eliminate it. Caution was needed, above all, with the cross section damming and congealing; this realization was patented in 1944 and has come to be known as the "area rule concept."

Other contemporary German designs, such as the Focke-Wulf Ta 183 or the DFS 346 research aircraft, were constructed with this particular tail-unit shape for the same reasons. Following the war, quite a few of the successful designs possessed this feature. The tests for this tail-unit shape began in the middle of November under Messerschmitt's contract with the Göttingen Aerodynamic Testing Laboratory (AVA). The six-part analysis, completed by the end of February 1945, was carried out by aerodynamicists with various horizontal tail-unit sizes and configurations on a full-scale model of the P1101.

With the 1101 experimental aircraft this horizontal tail-unit shape was to demonstrate its effectiveness.

During the swept-wing testing program, which

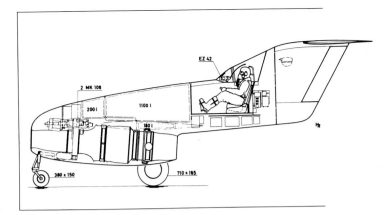

Long view of the P1106 design from December 12, 1944 - Sketch from H. Redemann.

Messerschmitt completed along with the DVL, it was recognized that a smaller aspect ratio could reduce resistance. (A realization whose accuracy demonstrated, among others, the delta wing and the Lockheed F-104 "Starfighter").

In comparison to the P1101, the wing surface remained virtually the same, the wing span was reduced by 1340mm and thereby lowered the aspect ratio of the P1106 to a value of l=3.5 (For a comparison: Me 262 l=7.05, P1101 at l =5 in October 1944).

Moreover, the deeper wing with similar profile thickness allowed for increased overall height.

On December 12, 1944, there was a further design for the P1106 which was extensively identical to the first studies from December 4; the differences were only noticeable when put under a microscope, as, for example, the fuselage shape or the design of the area in which the engine's blast exits.

As with many projects, there are a multitude of studies, derivatives, and variants. In order to prevent potential objections of the structural components being too small in size, and to meet possible continuing requirements of the RLM in regards to the fuel capacity, Messerschmitt's Project Bureau demonstrated that even this concept could have versatile applications.

In a variant from December 14, a "P1106 with increased operating range" was proposed. The construction showed a basic design with an increased wing span of 7.37 m, a lengthened fuselage, and therefore a possible fuel capacity of 2100 liters.

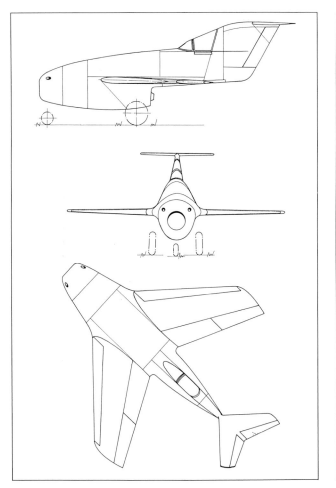

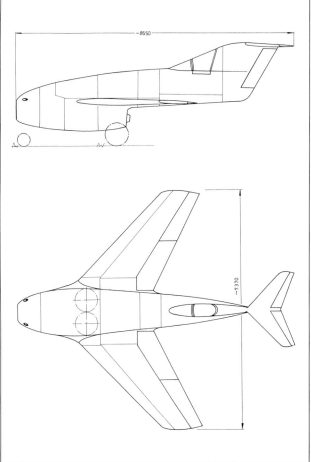

P1106 proposal from December 12, 1944 (left).

P1106 with increased operating range from December 14, 1944 (above).

Another extremely interesting derivative from the 1106 had the same date.

Messerschmitt planned to have an experimental aircraft for the P1106, just as with its predecessor, the P1101; to be sure, this would advance the speed programs, which, in all probability, would have been unattainable for the P1101-V1: supersonic speed.

How wholeheartedly the planners were with their thoughts as 1944 turned into 1945 is hardly worth tracing; the basic, formulated idea remained, and out of this the nose received, in place of the intake, a cone-shaped fairing, the T-shaped tail unit turned into a V-shaped tail unit, and above all the Heinkel jet engine was replaced by a Walter rocket engine.

Although to all concerned is must be clear, that at this time, due to the cost for such a program, it would no longer be possible. This study concerned a proposal which would have had a chance for reaching its goals through a virtual transformation. A short comparison with the first supersonic aircraft in aviation history should illustrate the following: The higher engine thrust of the Bell X-1 would surely make up for this through its lighter and smaller construction and the better aerodynamics of the P1106.

Following the war, the Americans tested the V-shaped tail unit which had quite often been critically reviewed by the DVL ("Disturbances on the V-shaped tail unit must be feared at Mach speeds"), and the swept wing on a similarly modified model of the Bell X-1 in the wind tunnels of the research center at Langley. Although the attempts proceeded successfully, the prototype successor, the Bell X-2, received a 40 degree swept wing, though only a conventional tail unit.

A further P1106 variant was made public in November, 1976 under the heading "Little known aircraft projects from the Second World War", the magazine "Luftfahrt International" (International Aviation).

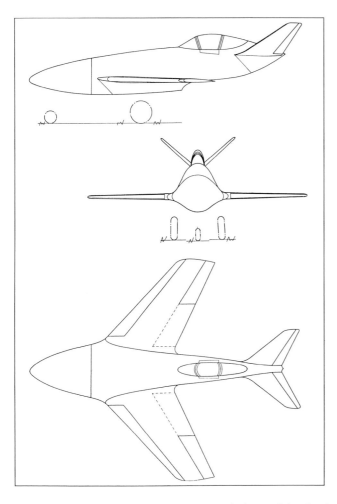

from the Messerschmitt construction (for example, the slots are missing);

• The size of the undercarriage's wheels are not recorded in any German tire charts;

• and, most importantly, the inserted engine is, to be sure, not of German construction from the time of the Second World War.

But, back to the actual 1106 main design: Messerschmitt invited discussion of the design from

Proposal for a supersonic test aircraft based on the P1106 with a rocket engine (left).

American high-speed, test aircraft Bell X-1; clearly recognizable is the fuselage shaped like a projectile.

According to the author's opinion, this design came from the time immediately following the war and was brought to paper in either England or France. The following accounts for this:

• The wing is, in its construction, totally opposite

December 12, 1944, in a somewhat modified form, at a comparison of fighter projects in Berlin-Adlershof from 19-21 December.

Similar to the rocket-driven 1106 studies from December 14, the project now had, instead of the T-shaped tail unit, a V-shaped one.

Unfortunately, from the proposal submitted, there were apparently neither specifications with a detailed survey diagram, nor bona fide performance ratings.

At the conference in Berlin an intense exchange of ideas took place between the firms concerned and the experts from the DVL. The resultant agreement of the comparative and common methods of performance calculation were to be discussed at the next meeting in January 1945. One consequence of this conference was that some firms had improved their designs or formulated new proposals. Messerschmitt did both: he retooled the P1106 and, in the first half of January, began work on the P1110.

P1106R in comparison:		
	Research aircraft Bell X-1	Research aircraft P1106 with rocket engine
Rocket motor thrust	2700 kp	approx. 1800 kp
Length	9500 mm	8420 mm
Height	3400 mm	3050 mm
Wing span	8500 mm	6740 mm
Wing surface	12 m2	13 m2
Wing thickness	10 % (2nd prototype)	12-8%
Spiking	0.5	0.4
Sweep	-	40 degrees
Aspect ratio	6.03	3.5
Take-off weight	6124 kg	approx. 4000 kg (Estimated with the Me 163 fuel capacity)

The survey diagram with the improved P1106 design was completed by Project Engineer Thieme immediately before the conference held at Berlin-Adlershof between 12-15 January. The aircraft represented in this diagram, along with the first 1110 design, was brought up for discussion by Messerschmitt with the DVL.

This P1106 was certainly criticized; this from an assessment made public to the aviation world in 1953 by Engineer Dietrich Fiecke, representative of the Henschel Firm. He had taken part in the Berlin-Adlershof conference as an advisor:

Technical data of the P1106 design from December 1944	
Application	single-seat turbo-jet engine fighter
Engine	1 x Heinkel He S011 with 1300 kp static thrust
Armament	2 x Mk 108 with 100 rounds each; approximately 500 kg external load
Wing span	6740 mm
Length	8000 mm (according to the diagram)
Height	3120 mm (from December 12, 1944)
Wing surface	13.0 m²
Aspect ratio	l 3.5
Spiking	t 0.4
Sweep	g 40 degrees
Profile	Root: National Advisory Committee for Aeronautics 0012 - 40 Wing nose point: NACA 0008 - 40
Communications Equipment	FuG 15, FuG 25, FuG 125
Fuel	1200 kg equal to approx. 1450 liters (1 = 0.83 - 0.85)
Empty weight	2538 kg
Flying weight	3958 kg
Undercarriage	Nose wheel 380 x 150, main wheel 740 x 210
Wing load	304 kg/m²
Thrust load	3 kg/kp
Performance characteristics (calculated)	
Maximum speed	1090 km/h at altitude of 6 km
Climbing speed	26.6 m/sec at sea level
Operating range	1600 km at altitude of 12,000 m
Take-off distance	620 m wit 500 kp additional thrust (booster rockets)
Landing speed	174 km/h
Landing distance	570 m
Operating ceiling	14,000 m
Structural design	full-metal fuselage wings made of wood

"The sweep shape and tapered form of the wings are a good selection, from a flow-technical standpoint. A leading edge flap is to be used as a landing aid. With a good physical layout, this leading edge flap also brings a considerable drag increment during high-speed flight, since the wings' profile would be detracted from at the most sensitive part of the upper, forward profile contour.

"The fuselage, with its cockpit set far to the rear of the aft fuselage, illustrates good aerodynamic form. The chosen engine enables a short air flow to the engine and an advantageous solution for the engine intake on the fuselage nose. Therefore, the loss of thrust is set at only 0.75% and has advantages for the engine. The design of the cockpit seems disadvantageous, since the wing is positioned right in the field of view.

"To reduce resistance, the horizontal tail and vertical tail were integrated into a large V-shape. At the root of both of these portions of the tail unit, pressure shock is to be expected as a result of the collapsing overspeed of the air, where, during high-speed flight, the tail unit's resistance is highly reduced.

"Right down to the tail unit's V-shape, the aircraft's contours are, from an aerodynamic standpoint, proportionately well chosen for obtaining high speeds, whereby the maximum speed, in relation to the other designs, remained quite high. On the other hand, due to the small wings and the large wing load, the data for take-off distance and landing speed are unfavorable."

The observations of the large wing load and the restricted field of vision for the pilot were topics that could hardly be altered without decidedly massive changes to the present concept. An improved field of vision is practically impossible without a miracle. At any rate, the Project Bureau reworked the proposal once again, fitted it with a new, large, 15.8 m2 wing assembly "A", raised the installed weaponry, refined the fuselage's contours, and equipped the aircraft with a normal tail unit.

In comparison with even the latest version of the P1101, this design maintains the upper hand and, together with the decidedly improved P1110 and radically new P1111 design, Messerschmitt submitted this to the RLM.

In comparison to the P1101, the P1106 design had, aside from the somewhat shorter engine intake, no advantages whatsoever. Thus, the imperfect full-view diagram from February 2, 1945 represents the P1106's finale. Construction of this design has not surfaced all too often since the war

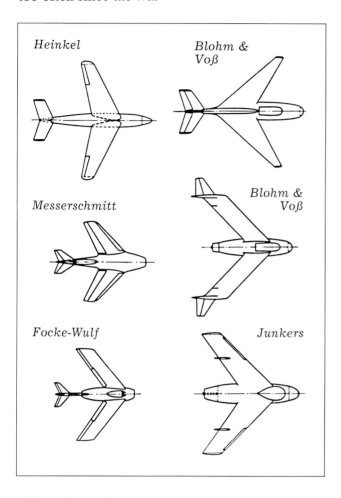

Heinkel

Blohm & Voß

Messerschmitt

Blohm & Voß

Focke-Wulf

Junkers

The six designs of Blohm & Voss, Focke-Wulf, Heinkel, Junkers, and Messerschmitt in plain view. These were the designs presented at the DVL conference in December in Berlin-Adlershof. Conference leader: Professor Dr. A.W. Quick. (above)

P1106 design from January 1945. (below)

There are certain similarities with the Italian light fighter Ambrosini "Sagittario II" from Sergio Stefanutti, who took part in the NATO competition for a light, fighter-bomber and who had the first Italian aircraft break the barrier to the speed of sound on December 4, 1956.

From the workings for the RLM competition and from Willy Messerschmitt's own method of construction, it clearly appears that the Project Bureau's concept, which was far ahead of its time, proved quite an adequate, forward-thinking model for light-fighter designs.

Technical Description
The P1106 design from January 1945:

The P1106 design is a direct derivative of the P1101. In comparison to the prototype, the modified configuration of the fuselage's main structural components stands out. With the configuration of the remaining components, the project leaders attempted to utilize parts with the smallest possible dimensions.

Located in the forward fuselage section are: a short air intake tube for the engine; both Mk 108's with 100 rounds each; and a hinged nose wheel (size 500 x 175 mm) that retracts to 90 degrees. The center fuselage section houses the large, 1350 liter, armor-protected fuel tank and the HeS 011 engine directly underneath. The wings' main spar runs between the fuel tank and the engine; in the bulge of the tank, immediately in front of this main spar, the forward-retracting main undercarriage with tires sized 740 x 210 mm finds its place. The partly armored cockpit lies immediately behind the fuel tank and the oval aft fuselage has sufficient space for housing the communications equipment.

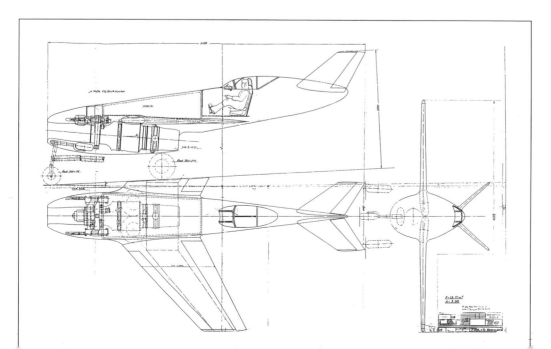

The tail unit's dihedral angle amounts to 50 degrees. With this design, the horizontal tail, as it were, takes on the functions of the vertical tail.

For lift increasing devices, the two-piece, swept, wooden monocoque wing possesses a slot on the wing's leading edge and, for flap ailerons, landing flaps. The wings carried approximately 145 kg of un-

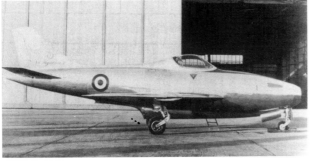

The P1106 design from February 22, 1945. Messerschmitt did not submit this proposal with the RLM or the DVL. He decided to submit the 1101, 1110, and 1111 projects.

Construction of the "Sagittario II" reminds one of the 1106 design from February 22, 1945. Since there is no valid data available from the "Sagittario II" design, a comparison with the RLM's proposal from January 1945 is impossible.

protected fuel.

For a comparison with the P1106, the following chart shows the available data of the Italian light fighter, Ambrosini "Sagittario II." This aircraft, which beat out the FIAT G-91 in NATO competition, is similar in many respects to the Messerschmitt proposal and thereby offered some revealing comparisons.

Without question, the "Sagittario", with its lower wing load, would be considered more versatile than the P1106.

The result: The P1110 Project - the "classic jet fighter" in post-war times:

The 1110 project arose in the first-half of January, 1945.

After Messerschmitt recognized that there was still more to be gleaned, considering the knowledge and experience gained during the past months with the 1101/1106 basic design, his Project Bureau began to develop an improved variant of the P1106; this resulted in a new aircraft.

Messerschmitt's project engineers attempted to reach the smallest possible fuselage cross-section by employing a configuration where the main compo-

nents of fuselage are set immediately behind one another. Now the ultimate cockpit size centered around the cross section of the main bulkhead. One difficulty surfaced at this time, and was dependent upon a satisfactory resolution of the design: The small fuselage diameter did not allow for nose intakes, and for air intake in the wing root, where the undercarriage is located, the wing was still too thin considering the contour that had been selected.

With the first design, the somewhat unusual form of the circular air intakes immediately behind the cockpit canopy was chosen. By using boundary layer suction in the intake area, the loss should be kept to the absolute minimum.

Messerschmitt submitted this proposal together with the improved P1106 to the RLM. In the already mentioned DVL conference in Berlin-Adlershof from 12 to 15 December, 1945, the submitted proposals were discussed (Blohm and Voss Bv P.212, Focke-Wulf Ta 183 I and II, Heinkel P1078, Junkers EF 128, Messerschmitt P1106 and P1110) and undertook similar performance ratings according to the already established methods.

	Messerschmitt P1106 January 1945 single-engine jet fighter, pilot in pressurized cockpit	Ambrosini "Sagittario II" single-engine jet fighter, pilot in pressurized cockpit with ejection seat
P1106 in comparison:		
Use		
Engine	1 x Heinkel HeS 109-011 A-0 with 1300 kp static thrust	1 x Rolls Royce Derwent 9 with 1633 kp static thrust
Dimensions		
Total length	9150 mm	approx. 9700 mm
Fuselage length	8650 mm	
Height approx. 3150 mm	3370 mm	
Wing span	6650 mm	7500 mm
Wing surface	13.17 m²	14.73 m²
Aspect ratio	= 3.36	= 3.82
Dihedral angle	0 degrees	0 degrees
Sweep t/4 line	40 degrees	40 degrees
Profile	Root NACA 0012-40 Tip NACA 0080-40	
Weight		
Empty weight	2300 kg	
Fuel	1200 kg	
Total load	1700 kg	
Take-off weight	4000 kg	3293 kg
Weaponry	2 x 30 mm Mk 108's with 100 rounds each and external load	2 x 30 mm Mk and external load
Maximum Surface load	303.7 kg/m²	224 kg/m²
Maximum Thrust load	3.08 kg/kp	2.02 kg/kp
Performance		
Maximum speed	993 km/h (M 0.882 at altitude of 7 km)	1006 km/h in nose dive Mach 1
Taxiing distance During start	830 m	
Initial climbing speed	21.2 m/second	
Service ceiling	13,300 m	14,000 m
Landing speed	184 km/h	
Maiden flight	End of 1944/beginning of 1945	Predecessor: January 5, 1953 Sagittario II: May 19, 1956

Apart from the fact that the P1110 exhibited the highest speed of all designs submitted, it was overlooked and criticized, just as the remaining designs had been.

In detail, the DVL found the following faults:

- The leading edge flaps; the fears are easy to understand, since Messerschmitt overlooked the two fastest German fighter aircraft with the same feature, the Me 262 and the Me 163;
- the high wing load, which caused inadequate maneuverability in curves and poor low-speed flight characteristics;
- the annular air intake:

"the air intake to the engine, in the fuselage midsection, seems to not have been an advantageous choice, since the air intake lies in the fuselage's largest area where pressure shock appears during high-speed flight. The pressure shock inhibits a clear flow of air to the engine" and "the flexible shaping of the fuselage, which already exists in the annular engine intake of the 1110, causes unfavorable after-effects."

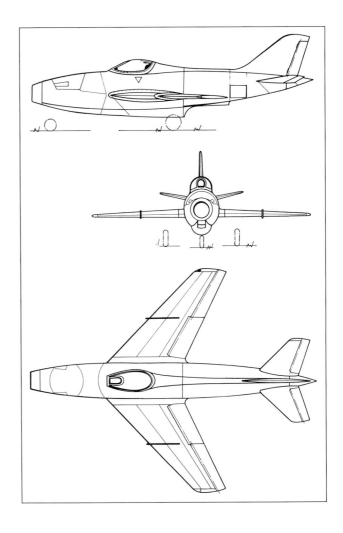

For the "intake shock", the following is to be noted: The pressure shock during high-speed flight occurs with the already patented area rule concept on the largest cross-section. With the essential inclusion of the wing assembly, the site of the largest cross-section lies behind the annular air intake!

Technical data of the P1110 design from January 12, 1945: The data is compared to that of a light fighter which belonged to the competitors of the NATO bidding war from early 1954 and was built in large numbers: the Folland Fo. 141 "Gnat."

This comparison should illustrate how far Messerschmitt was ahead of the pack with his concept of 1944/1945.

The flight performance of the Me P1110 was calculated to have an intake thrust loss of 8% of the engine output.

The RLM still could not relinquish a contract for construction of mass-produced aircraft despite the criticism from the DVL. There were still thoughts, however, concerning the principle process:

A hand-written draft from Messerschmitt to Voigt and Hornung; Messerschmitt immediately recognized the weakness and discussed the intake problems surrounding an engine placed in the rear of the fuselage. (January 4, 1945)

P1110 in comparison:	Messerschmitt P1110 design January 12, 1945	Folland (Hawker-Siddeley) Fo 141 "Gnat" F.1 variant
Designation	single-seat fighter aircraft	single-seat small fighter
Development	1944/1945	Development began in 1950, maiden flight of the proto type Fo 139 "Midge", August 8, 1954
Crew	1 pilot in pressurized cockpit with ejection seat	1 pilot in pressurized cockpit with ejection seat
Engine	1 x Heinkel He S011 A with 1300 kp static thrust	1 x Rolls-Royce (Bristol) "Orpheus" with 2050 kp static thrust
Wing span	6650 mm	6750 mm
Sweep	40 degrees	40 degrees
Aspect ratio	3.36	3.58
Relative wing area	12%	8%
Wing surface	13.17 m²	12.7 m²
Length	9665 mm	9070 mm
Height	approx. 2700 mm	2690 mm
Empty weight	2580 kg	2200 kg
Take-off weight	4000 kg	4030 kg
Sheer load	3.07 kg/kp	1.97 kg/kp
Maximum wing load	304 kg/m²	317 kg/m²
Maximum speed horizontal flight	1006 km/h in 7,000 m altitude	approx. 1150 Mach 1.0
Initial climbing speed	26 m/second	101.6 m/second
Service ceiling	13,100 m	approx. 15,000 m
Operating range	1500 km	1900 km
Weaponry	3 x 30 mm Mk 108 and 500 kg of (German Mk 213C)	2 x 30 mm Aden Mk jettisonable armament and 454 kg jettisonable armament

The other firms likewise attempted flight testing and to corroborate their theoretical work.

In the middle of February, Blohm and Voss received instructions from the RLM to investigate swept wing construction and tail unit's configuration of the BV P.212 in preliminary tests for their rigidity and flutter stability. Focke-Wulf had already begun construction of a Ta 183 test bed, which, as in the case of the P1101, was to be test flown with a Junkers 004 turbine engine.

Messerschmitt received the recommendation from the DVL for further enlargement of the P1110 design's wing surface and to retool the engine intake.

After the wing surface enlargement only brought a relatively minor drag increment, even through significant improvements in the area of low-speed flight characteristics, the proposal for Messerschmitt was acceptable. (The DVL proposed not having wing surfaces under 20 m2, Messerschmitt raised the P1110's to 15.8 m2 and drafted the tailless P1111 design with 28 m2.) Furthermore, the DVL expressed doubt concerning the V-shaped tail unit: "It appears questionable whether the shape of the tail unit, chosen for its low resistance, has such disadvantages with the lower surface size that such a requirement cannot be replaced."

"The wing sweep necessary for high-speed flight brings a long list of difficulties in the area of flight mechanics and can only be overcome, in part, through testing in high-speed flight wind tunnels and through actual flight."

While wind tunnel testing was carried out at various, overburdened research installations, and led to useful results, things weren't favorable for actual flight testing, given the lost cause at hand.

During this time frame, Messerschmitt was the only one with a test bed in the final assembly stage and waited, in vain, for the required engine; furthermore, a similarly retooled Me 262 would be brought in for swept wing testing (if the war situation would still allow).

Although the doubts were seemingly natural and, following the war, were not to be confirmed through flight tests (with the Republic XF-91 and the Supermarine Type 508, for example), Messerschmitt decided first to use a normal vertical tail with a horizontal tail unit.

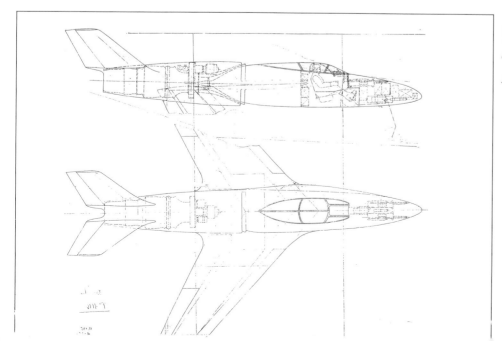

The Folland Fo-139 "Midge" (prototype of the "Gnat"). The aircraft is similar to the P1110 in its dimensions and configuration. (above)

First design of the P1110 from January 1945. (left)

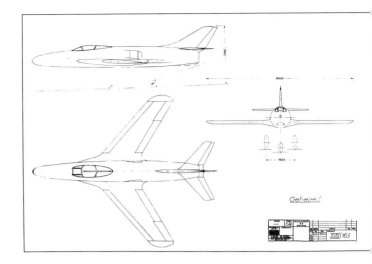

P1110 proposal from February 2, 1945.

First use of the "Wing assembly A", which was finally detailed in the second half of 1945. The drawn-in inner wing is relevant even today (on the F-18 or MiG-29, for example). This aerodynamic twist produces strong vortex flow on the wings' topsides. The migration of the boundary layer is thereby prevented to a certain degree -flight with a larger angle of incidence is then possible.

P1110 proposal from February 12, 1945 (middle).

Sketch for the new P1110 design (probably from Hornung immediately after the end of the war).

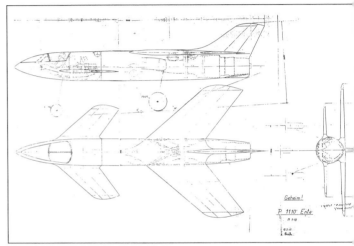

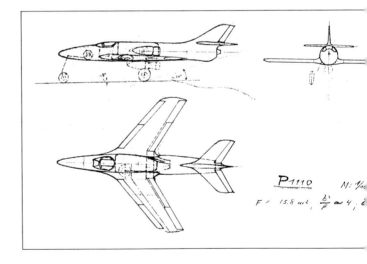

As early as the beginning of February a full-scale design showed extensive improvement in the P1110 (Designation number XVIII/165 from February 2, 1945). This analysis shows, among others, the new enlarged wing assembly with modified contours, lateral air intakes, and a conventional tail unit. The internal armament remained unaltered by using 3 x Mk 108's.

At about the same time, the Hornung Project Group was busy with modification of an alternative that allowed for a wide range of speeds. Short take-off and landing distances were, right next to high-speed flight, a matter of survival.

The planned separated or unseparated tiltable wing, already undergoing multiple studies, did not come into the picture due to the complicated mechanics involved; therefore the project engineers were looking into studying a tailless design with a low wing load (P1111), in which they proposed the so-called "duck construction" for the P1110. This should permit good pitch and lateral stability on the aircraft for low-speed flight characteristics. A further advantage of this construction type is the possibility of compensating larger displacement without a noticeable loss to aerodynamic quality. With no further experiments in the picture, Messerschmitt dropped the proposal in favor of the "conventional" P1110.

Certainly the specialists at the DVL and the RLM would have expressed fears that the "duck construction" had its own directional stability problems. It is interesting in this context to note that, in 1946/1947, the design team of the Swedish team from the SAAB Firm, in possession of the Messerschmitt documents, proposed the same "duck construction" during the groundwork for the J-32 "Lansen" fighter bomber.

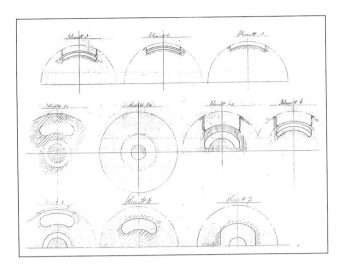

Intake problems of the 1110: Project engineer Thieme completed this design of the intake section for construction of a wind tunnel model.

The Swedish Air Force showed little inclination of accepting such a design and so the J-32 received the appearance of the "conventional" P1110 design.

Today, many of the most modern combat aircraft designs show a more or less distinctive "duck construction."

Messerschmitt and his coworkers were united in the middle of February concerning the proceedings, and the Project Bureau was to formulate a detailed, full-view design. Troubles ensued, however, just as with the P1101, with the housing for the undercarriage and the shaping of the air intakes.

The successful resolution of the intake problem was significant and, accordingly at this juncture, the engineers exercised extreme care.

As has already been noted, the two annular intakes lateral to the fuselage surface were no longer a consideration. There were not many possibilities remaining without abandoning the design philosophy of area rule concept for fighter aircraft: air scoops, whether lateral or placed above, would have enlarged the cross section at this sensitive location, and moving the air intakes forward, just as with friction loss, would have meant turning away from the principle of the smallest possible fuselage.

In order to restrict intake loss, on the other hand, suctioning the incoming boundary layer was planned for the first designs in January. Wind tunnel experiments with an air intake half the size illustrated the effectiveness of the air flow and the boundary layer control system. In comparison to the preferred nose or wing root intake, this solution was clearly not as adequate. By using boundary layer suction and, above all, due to the excellent aerodynamic lay-out of the fighter, this disadvantage could be more than counterbalanced: the calculated maximum flight speed of the P1110 was, according to the recent comparison of the aircraft on February 27/28 February, 1945, in Oberammergau, just over the Messerschmitt P1111, and thereby the fastest aircraft among the submitted designs. The P1110 received a favorable assessment in regard to its expected performance and flight characteristics. The engine's air intake received a bad rating, just as the preceding P1110 design had, and the selected arrangement was noted as having an instability factor on its performance rating.

With the rapid technical progress and, above all, due to the catastrophic military-economical situation, the technical shortcomings had to be almost continuously adjusted for: therefore the Special Commission Day Fighter, chaired by Willy Messerschmitt, proposed new, expanded "technical guidelines" for the standard aircraft. After this conference, in which no decision was made concerning construction of a mass-produced aircraft, the leaders of Messerschmitt's Firm prepared for further proceedings. What was clear was that, at this point in time, the simultaneous development of three high-performance would have been neither sensible nor possible. With the knowledge gained thus far, there were two directions one could take for the best possible success:

- further development of the "conventional" concept (P1101, P1110), for which a test aircraft would shortly be completed;
- or the realization of the tailless design (P1111).

A third possibility began by combing through both of the above concepts. Although no final decisions were made, the technical direction of the project took this advantage during March.

Following the war, Woldemar Voigt formulated the goal opposite the Americans as follows: "Within one year, we intended, as a first step of our research work, to produce a high-speed aircraft which will surely reach a speed of 1000 km/h."

In the middle of March, following studies, the Project Bureau produced the most diverse weapon mountings (Mk 108, Mk 112, and Mk 214) under design P1110 W, and the fuselage nose for this designator already possesses the faired cockpit for the P1112. One of these designators (P1110 page 2 with 4 x Mk 108 from March 14, 1945) will then be used, logically, to provide a replacement for a P1112-V1 mock-up.

Work on the P1110 design, exactly like the P1111 proposal, ran into Woldemar Voigt's high-speed aircraft project, designated P1112-V1.

Nevertheless, Messerschmitt's actual P1110 concept, just as the P1101, was not useful due to the already mentioned reasons, though jet aircraft of similar configuration crowded the skies following the war.

Many of those aircraft are named here:
- Saab J-32 "Lansen"
- Folland "Gnat"
- Dessault "Etendard"
- Hawker "Hunter"
- Supermarine "Swift"
- Grumman F-11 "Tiger"

Technical Description: The P1110 Project: The revolutionary design, the basis of the P1101, became one of the two prototypes for the construction and testing of the planned P1112-V1 immediately following the war.

The main components of the P1110 were configured according to Hertel, Frenzel, and Hempel (Junkers) area rule concept from March, 1944.

The following description is taken from Technical Report TB 139/45 "Comparison of the Single-Engine Jet Fighters Submitted." The authors of this report from February 26, 1945, are Hans Hornung and Woldemar Voigt.

Specifications of the P1110 Project of the Messerschmitt Firm AG:

In general: Continuous winged, low-wing monoplane with engine installed in the rear and a normal tail unit assembly configuration

Wing: Similar to the P1101 series

Fuselage: Circular fuselage made of all metal construction. Fuselage nose is composed of a nose cone for a weapons turret with 3 x Mk 108, and the nose

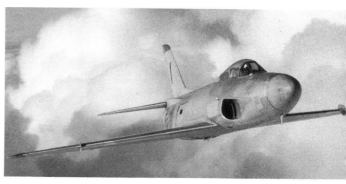

Saab J-32 "Lansen"

wheel space, connected to the pressurized cockpit with the communications equipment. The fuselage center section consists of the wing, the main undercarriage and the armored fuel tank. And the intakes with the boundary layer suction system. Included within the rear fuselage section are the engine and tail unit.

Tail unit: Normal four-blade tail unit made of wood, with a rear sweep of 40 degrees. Can later be replaced by a V-shaped tail unit.

Under-Nose undercarriage unit is a single-strut undercarriage with offset wheel carriage: fork, is retracted into the weapons turret with a hydraulic strut. Nose wheel size - 500 x 180. The main undercarriage is retracted in a forward direction into the wing-fuselage area by a self-latched hydraulic strut, and is rotatable around an axle.

Engine: The engine lies in the rear fuselage. The air is directed into the engine through a hole on either side of the bottom fuselage. The boundary layer, running along the length of the fuselage nose, is sucked by a compressor attached to the engine via at least two, and maybe several, slots in the bent intake.

Fuel system: 1500 liter armored fuel tank in the center fuselage section, between the cockpit and the engine. The planned center tank can be exchanged with a normal cell-type tank.

Equipment: Standard equipment for a fighter aircraft.

Weaponry: 3 x Mk 108 with 2 x 70 and 1 x 100 rounds in weapons turret. Installation of two additional Mk 108's is possible.

Armor: Pilot armored against rounds of 12.7 from the front and 20 mm from the rear.

Technical Data:
Primary Function:
light air superiority fighter
Crew:
1 pilot in a partly armored, pressurized cockpit with a 99 mm thick windshield which is inclined 30 degrees from the body axis.

Engine system:
1 x Heinkel HeS 109-011 A-0 static thrust 1300 kp with a compressor sitting on the engine shaft for taking in the boundary layer from the lateral air intakes.
Compression rating: 200 PS corresponding to 4 % of the engine thrust.
Later: engine with 1 x HeS 109-011B with 1500 kp static thrust.

Engine installation in the rear fuselage should compensate for the weight of the weaponry in the fuselage nose. During weapons firing, the center of gravity displacement is compensated for with the weapons discharge. Despite the somewhat inconvenient air intakes, the drag coefficient of the P1110 is 16 % lower than that of the former 1101 and 1106 projects.

Main data:
Total length:	10360 mm
Total height:	3180 mm
Wing span:	8250 mm

Fuselage:
Fuselage length to nozzle:	9350 mm
Highest fuselage height (with cockpit):	1200 mm
Fuselage diameter behind cockpit:	1120 mm

Wing assembly:
See P1101 series for this data

Undercarriage:
Nose whee	1500 x 180 mm
Main whee	1 740 x 210 mm
Wheel base	1650 mm

(from the February 22, 1945 layout)

In the end, the Project Bureau leaned toward an inward-retractable main undercarriage with a 2450 mm wheel base; the nose wheel would then extend forward toward the fuselage nose. These came from the design depicted and from the article "New Designs" in the Woods Report of U.S. specialist (civilian) J.M. Schoemaker.

Tail unit:
Vertical tail surface (with rudder)
1.8 m^2
Horizontal tail surface (with rudder)
3.0 m^2

Weights:
(According to H. Hornung from June 27, 1945)
Wings	450 kg
Fuselage	350 kg
Tail unit with controls	135 kg
Undercarriage	230 kg
Aerostructure (total)	1165 kg
Engine with components	1015 kg
Equipment	190 kg
Weaponry (Mk 108 and armor)	442 kg
Empty weight	2812 kg
Pilot	100 kg
Fuel (1500 liters	1250 kg
Ammunition	128 kg
Removable load	1478 kg
Normal take-off weight,	4290 kg

without additional equipment and armament (i.e. with standard equipment only)

Flight performance:
Maximum speed	
at sea level	902 km/h
at altitude of 7 km	1015 km/h
Climbing speed	
at sea level	22.2 m/sec

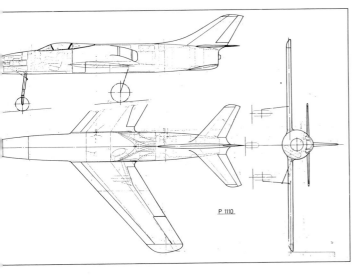

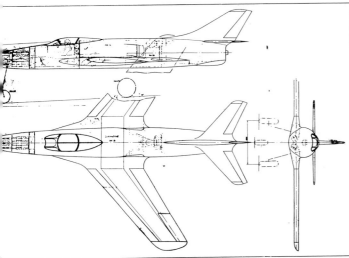

P1110 with modified undercarriage: layout not totally completed nor dated, probably the first proposal of the "real" P1110 with the "Wing assembly A." The project engineers combined the P1110 and the P1111 proposals for the P1112.

P1110 proposal, which was discussed with the DVL on February 27 and 28; also visible is the modern, wide wing root.

Service ceiling	14,000 m
Taxiing distance during take-off without boosters	approx. 800 m
Landing distance	approx.180 km/h
Stalling speed	160 km/h
Taxiing distance during landing	610 m

The following data came from an American CIOS (Combined Intelligence Objective Sub-Committee) report:

Service ceiling	greater than 12,000 m
Flight time at sea level	40 min
Flight time at 10 km altitude	1,8 hours
Operating range at sea level	500 km
Operating range at 10 km altitude	1500 km

Performance comparison:

Although Professor Messerschmitt's colleague, Gabrielli, an engineer, designed the Fiat G-91 for a different purpose, there was obviously closely linked data between the victor of the NATO light fighter competition from 1954 and the German fighter design from 1945. Once more, the German aircraft designers were far ahead of the pack.

The Alternative: P1111/P1112

P1111: After both Messerschmitt proposals were released at the DVL conference in Berlin-Adlershof, Messerschmitt decided upon the following: the P1101 production series design received the newest technical standard equipment; conventionally designed P1110 proposal would be revised; and the 1106 proposal would be dropped: and from here a totally new concept would emerge.

In January, 1945, the Project Bureau began work on this new aircraft, designated P1111. Using this design, the engineers circumvented the difficulties with

Construction of the Fiat G-91 in an earlier portrayal. The P1110 configuration is clearly visible in the main components. (on the right)

Parachute landing of a Fiat G-91 on a grass strip. (below)

the air intakes, the instability in the construction of the tail unit, and the terrible and much complained about low-speed flight characteristics to design an extremely refined aircraft. The design was without a horizontal tail and possessed a highly swept wing which boasted a large depth at the wing root. The profile attained through this allowed the problem-free configuration of the air intake at the wing assembly root and also permitted accommodation of a wide wheel base for the undercarriage, whose wheels retracted into the wing-fuselage connection.

Messerschmitt had a lot of experience with aircraft of this construction type due to the design of the Me 163, and both designs can hardly be compared, though not only because of the totally diverse engine systems. The table on page 94 should clearly show the aerodynamic differences, and then shine light on the P1111 concept.

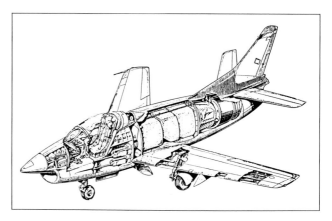

Within only one month, this new concept took on concrete form. About one week before the conference deadline, the Project Bureau constructed a detailed full-view sketch in the scale of 1:10 and the complete performance calculations, based on the method established in January, could be provided.

The one item that did not remain was the entire work up of the engine; it was planned to house the entire fuel capacity in unarmored tanks in the outerwing.

On the 24th and 25th of February, a two-seat, twin engine jet night fighter was derived from the already available design. The night fighter was designated Project Number 1112. It is possible that this study was used as only for the purpose of a comparison with the developed night fighter proposals from the Me 262. In any case, the P1112 night fighter was hot off the press, and the project number could be used as a continuation in the development of the P1111 design.

Three-sided view of the Italian light fighter and strike aircraft. (from "Aeromodeller", January 1959).

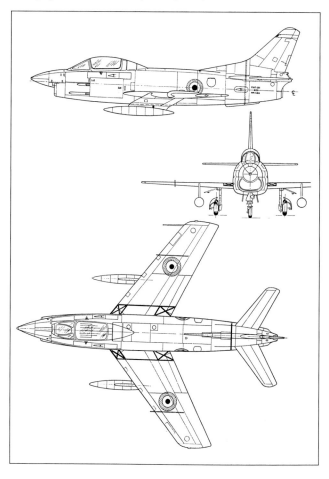

P1110 in comparison:		
	Messerschmitt P1110	Fiat G-91
Application	air superiority attack aircraft	strike air craft, low-level
Engine	1 x HeS 011 with 1300 kp static thrust	1 x Fiat 4023.002 (License BS "Orpheus" 80302 with 2270 kp static thrust
Length	10,360 mm	10,290 mm
Height	3180 mm	4000 mm
Wing span	8250 mm	8560 mm
Wing surface	15,85 m²	16.42 m²
Aspect ratio	4.29	4.46
Sweep t/4-line	40 degrees	37 degrees 13' 24"
Wing assembly profile Interior	NACA 0008-40 (8%)	NACA 65A 112 (12%)
Exterior	NACA 0012-40 (12%)	NACA 65A 111 (11%)
Aerodynamic aids on wing assembly	Leading edge flaps, landing flaps, geometric dÈcalage	boundary layer fence, aerodynamic dÈcalage
Empty weight	2812 kg	3040 kg
Additional weight	1470 kg	2162 kg
Take-off weight normal	4290 kg	5202 kg
Landing weight Maximum	3490 kg	3730 kg
Fuel	1200 kg	1675 kg
Wing load	271 kg/m²	317 kg/m²
Power loading	3.3 kg/kp	2.3 kg/kp
Armament	3 x Mk 108 with 2 x 70 rounds and 1 x 100 rounds and approx 500 kg external load or 5 x Mk 108	4 x 0.5' Colt Browning or 2 x 20 Mk or 2 x 30 mm Mk with approx 500 kg external load
Horizontal flight speed	902 km/h-at sea level 1015 km/h-at 7 km altitude	1045 km/h at 1.5 km altitude
Initial climbing speed	21.5 m/sec	30 m/sec
Operational range (without supplementary tanks)	1500 km	2200 km
Landing speed	178 km/h	275 km/h
Minimum speed		230 km/h
Take-off taxiing distance	790 m	approx 900m
Landing taxiing distance	610 m	approx 300 m
Service ceiling	approx 14,000 m	12,200 m
As of:	February/March 1945	1960/61

P1110 in comparison:

	Messerschmitt Me 163-B	Messerschmitt P1110 Project
Length	5920 mm	8920 mm
Wing span	9300 mm	9160 mm
Height	2490 mm	3060 mm
Engine thrust	1 x 1500 kp	1 x 1300 kp
Armament	2 x Mk 108 with 60 rounds each	4 x Mk 108 with 100 rounds each
Undercarriage	droppable, three-legged skis	retractable, landing on undercarriage
Weights		
Empty weight	1775 kg	2740 kg
Flying weight	3945 kg	4282 kg
Maximum wing loading	210 kg/m²	153 kg/m²
Performance		
Maximum speed	950 km/h (estimated)	995 km/h
Service ceiling	14,500 m	14,000 m
Take-off distance	approx 800 m	600 m
Landing distance	approx 600 m	450 m
Geometrical data on wing assembly		
Wing span	9300 m	9160 mm
Wing surface	19.7 m²	28.0 m²
Sweep t/4	27.5 degrees	45 degrees
Lift increasing devices	slots, solid	slots
Aspect ratio	4.4	3.0
Spiking	0.42	0.3
Root edge	Max. thickness 14.4% in 30% rearward position	Max. thickness 8% in 40% rearward position
Wing tip	Max. thickness 8% in 20% rearward position	Max. thickness 8% in 40% rearward position symmetrical profiling
Decalage	5.7 degrees	
Angle of incidence from fuselage axis	3 degrees	
	Wing assembly unsuitable for trans-sonic flight	Wing assembly allow for trans-sonic and super sonic flight with comparable engine power

At the conference in Oberammergau on 27 and 28 February, just prior to the end of the war, during a design comparison, the Messerschmitt P1111 fared superlatively in the area of performance. To be sure, in the criticism of the fuel system of the designs in question, it was said: "The P1111 is further ruled out, in our opinion, due to the impossibility of protecting the fuel."

On the other hand, the remarkably good performance of the design would be emphasized at the conclusion. The outstanding areas were, above all, the combination of: high-speed flight capability, the take-off, climb, and landing performance. The experts held the opinion that these qualities demonstrate a fundamental superiority of the flying wing construction type.

Following the war, the P1111 concept had many imitators. The resistance lacking configuration of the main components, in conjunction with a highly swept wing without a horizontal tail, culminate into a most successful combat aircraft.

To prove this, one need only look at the German-inspired American navy fighter Douglas F4D-1 "Skyray" and the Swedish SAAB J-35 "Draken", whose beginnings lay in German documents.

The British research aircraft, de Havilland D.H. 108 possesses a direct resemblance to the Messerschmitt proposal. When comparing the concepts (see the Technical Descriptions), one is tempted to maintain that the deciding point from which the emulator differs from the original concept, was the reason for the tremendous difficulties during flight testing: had the wing assembly possessed a smaller aspect ratio similar to that of the P1111, the D.H. 108 would certainly have been easily in the high-speed flight ranges.

Despite this, the D.H. 108 is the first turbo-jet engine aircraft that broke the sonic barrier, and the test flights of the three prototypes delivered valuable data for development of the D.H. 106 "Comet" and the D.H. 110 "Sea Vixen." With the latter of the two, the navy fighter's aspect ratio was 3.86 (instead of 4.64, as with the D.H. 108), and the load reduction of the wings served to produce a horizontal tail with a doubled vertical tail.

Technical Description: The 1111 Project: The specification and the larger portion of data for Messerschmitt's tailless alternative to the 1101/1110 class originated from the already mentioned "Technical Report TB 139/45" from February 26, 1945.

B8 specifications of the Messerschmitt Firm AG

P1111

In general:	Swept, tailless mid-wing monoplane
Wings:	Wings configured to take on 1500 liters of unprotected fuel. Large sweep with smaller aspect ratio and spike. Slots are installed in the area of the aileron at the wing nose
Fuselage:	All-metal fuselage. Forward fuselage section is set from the nose cone, configured as a weapons turret with 2 x Mk 108 and a nose wheel area connected to the pressurized cockpit with communications equipment behind the pilot.
Tail unit:	Aileron works simultaneously as a horizontal tail. There is a standard swept vertical tail at the aft fuselage. Undercarriage: Nose wheel undercarriage is a single-strut unit,is drawn to the rear into the weapons turret. The wide-track main undercarriage is drawn laterally into the wing-fuselage junction. Nose wheel 465 x 165. Main wheel 740 x 210.
Engine:	Air flows from the engine through the two lightly curved valves which possessed openings in the wing-nose of the wing-fuselage junction.
Fuel system:	All fuel, 1500 liters, housed unprotected in the wing.
Equipment:	Standard equipment for fighter aircraft.
Weaponry:	2 x Mk 108 in the weapons turret with 100 rounds each. 2 x Mk 108 in the wing-fuselage junction with 100 rounds each.
Armor:	Aircraft is protected from the front against 12.7 mm and 20 mm from the rear.

P1111
Technical data:

Application:	light, air-superiority fighter
Crew:	1 pilot in pressurized cockpit ejection seat available
Engine:	1 x Heinkel 109-011 A-0 with 1300 kp static thrust
Main data:	
Total length	8920 mm
Total height	3060 mm
Wing span	9160 mm
Fuselage:	
Fuselage height without cockpit	1080 mm
Fuselage width	950 mm
Wing assembly:	
Wing span	9160 mm
Wing surface	28.0 m²
Aspect ratio	3.0
Sweep	45 degrees in 0.41 t
Interior profile	NACA 0008-40
Exterior profile	NACA 0008-40
Spiking	0.3
Undercarriage:	
Nosewheel	464 x 165 mm
Main wheel	740 x 210 mm
Wheel base	3130 mm
Vertical tail:	
Surface with rudder	2.0 m²
Weights:	(according to H. Hornung from June 27, 1945)
Wings	570 kg
Fuselage	200 kg
Tail unit	50 kg
Control	40 kg
Undercarriage	230 kg
Total aerostructure	1090 kg
Engine, with components	940 kg
Equipment	190 kg
Weaponry (Mk 108 and armour)	520 kg
Empty weight	2740 kg
Empty weight	2740 kg
Pilot	100 kg
Fuel	1250 kg
Munitions	192 kg
Additional load	1542 kg
Take-off weight	4282 kg
Normal equipment, without additional equipment or weaponry	
Landing weight	3482 kg
Flight performance:	
Maximum speed at sea level	900 km/h
at 7 km altitude	995 km/h
Climbing speed at sea level	23.7 m/sec
Taxiing distance during take-off	600 m
Landing speed	155 km/h
Taxiing distance during landing	450 m
Service ceiling	up to 14,000 m
The following data originated from an American CIOS report:	
Flight time at sea level:	40 min
Flight time at 10 km altitude:	1.8 hours
Operating range at sea level:	500 km
Operating range at 10 km altitude:	1500 km

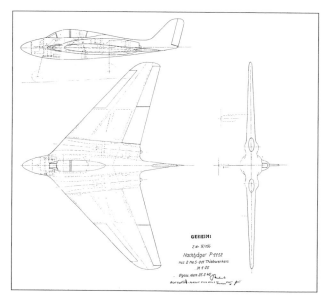

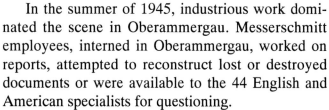

In the summer of 1945, industrious work dominated the scene in Oberammergau. Messerschmitt employees, interned in Oberammergau, worked on reports, attempted to reconstruct lost or destroyed documents or were available to the 44 English and American specialists for questioning.

Two British gentlemen, R.G. Bishop and R.M. Clarkson from the CIOS (Combined Intelligence Objective Sub-Committee) attended to the design basis of the planned, new generation of Messerschmitt aircraft. Both were employed by the English aircraft company, de Havilland; Chief Engineer Bishop is the creator of many successful aircraft.

Shortly thereafter, de Havilland received a contract from the Ministry of Supply (MoS), with specifications E. 18/45 and E. 28/45, to construct three research aircraft based on the D.H. 100 "Vampire", and to make the new, swept wing a reality for testing. The intention behind this was the creation of a new wing assembly concept for future transport and combat aircraft. Bishop turned his experiences in Oberammergau into reality and his D.H. 108 "Swallow" leaned closely toward the P1111, whose progressive concept had persuaded him.

The D.H. 108 was not considered a total success, since all three prototypes crashed. Nevertheless, the same data was used, to a large extent, in the construction of the D.H. 110 "Sea Vixen" and in the developmental work on the world's first jetliner, the D.H. 106 "Comet." Following the experiences with the D.H. 108, the aircraft received a horizontal stabilizer, though it was originally to have been fitted with only a vertical tail.

Basic design of a variation of the P1111: twin-engine night fighter P1112. (above left).

Messerschmitt Me-163: The aileron configuration and the rigid leading edge are easily noticeable. Messerschmitt retained this principle configuration for construction of the P1111. (above)

The English P1111: D.H. 108 "Swallow"; this photo shows the third prototype. (below)

Full-view drawing of the P1111 XVIII/168 from February 24, 1945. Arranger: Project engineer Scharrer. (right)

Messerschmitt carried out this step in March 1945, when he equipped the P1112, further developed from the P1111, with a V-shaped tail unit.

P1112 - Unfinished: Immediately after the experts expressed fundamental doubts about the P1111 proposal, Hans Hornung's Project Group once again began work on the single jet engine fighter in order to eliminate the known shortcomings.

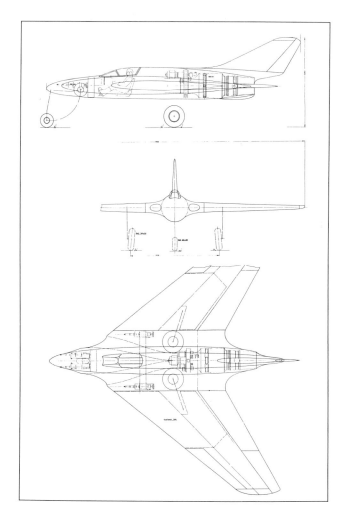

with his Technical Operations Staff, apparently decided to carry on with only what remained: a design. For this, the preferences of both lines of development were to be unified into one design. This illustrated a new, incompletely constructed design with the designator P1112/S2 and it was a new aircraft in many respects. Noticeably, the configuration of the air intakes is similar to those of the P1110: they are immediately in front of the engine, and the larger aspect ratio of the wing assembly and the reduction of the wing sweep from 45 to 40 degrees.

The P1112 S/2 design from March 27, 1945, almost identical to this conception, shows a further swept air intake, which appeared more often following the war as the NACA inclined intake, on, for example, the planned and already in prototype form F-86, successor from North-American YF-93, and the German Federal Republic's DFS-582, high-altitude aircraft design. The platform intake compensates for form drag

The technical framework for retooling formed new requirements of the "Special Commission Day Fighter" at the beginning of March.

A full-view drawing from March 3, 1945, under the title "P1112 S/2" shows an aircraft whose construction closely resembles that of its predecessor, the P1111.

The forward section of the somewhat smaller fuselage forms a completely faired cockpit. In the reduced wing assembly (22 m2* instead of 28 m2*), only a portion of the entire 1550 kg fuel capacity is housed in self-sealing tanks. Because of the reduced empty weight, it was possible to carry along 1900 liters of fuel without any weight disadvantages.

Sometime during the first half of the month of March when, in what remained of Germany and only chaos was prevalent, Professor Messerschmitt, along

P1111 in comparison:		
	Messerschmitt P1111	de Havilland D.H. 108, 2nd prototype
Application	turbo-jet engine fighter aircraft	swept wing research aircraft
Engine	1 x HeS 011 with 1300 kp thrust	1 x Goblin 3 with 1497 kp static thrust
Length	8920 mm	7470 mm
Height	3060 mm	approx. 3400 mm
Wing span	9160 mm	11,890 mm
Wing surface	28.0 m²	30.47 m²
Relative thickness	8%	11%
Sweep	45° in 0.41 t	45°
Aspect ratio	3.0	4.64
Angular setting to fuselage	0°	0°
Dihedral angle	0°	0°
Decalage	none	none
Aerodynamic aids	slots in the area of the aileron	slots in the area aileron
Take-off weight	4282 kg	4064 kg
Wing loading	153 kg/m²	133 kg/m²
Performance		
Maximum speed	995 km/h	1030 km/h
Landing speed	155 km/h	140 km/h
Service ceiling	14,000 m	12,200 m
Designed:	Jan/Feb 1945	Design began Oct., 194 Maiden flight: May 15, 1946. 2nd prototype: Aug 23, 1946

of the intake scoops and certainly represents the resistance-free construction of a lateral, fuselage-located, intake for flight in the supersonic range.

Therefore, the last fighter design which emerged in the Oberammergau Project Bureau also shows this unusual intake geometry. Otherwise, this aircraft represented here and designated P1112-V1 from March 30, 1945, differs in some fundamental areas from its predecessors:

- The new use of the V-shaped tail unit stands out upon first look; through close scrutiny of the original designator XVIII/166, it is plain to see that Project Engineer Scharrer also evaluated the aircraft's equipment with the "normal tail unit" of the P1110 design from February, 1945

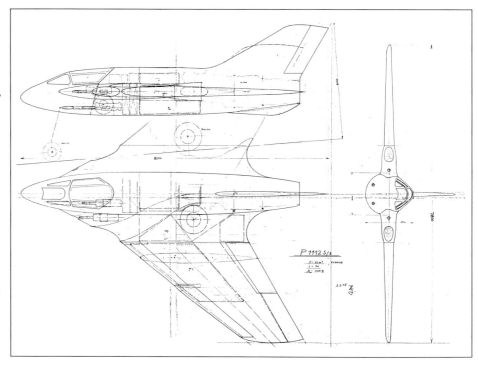

Full-view sketch of the P1112 S/2 from March 3, 1945: Arranged by Hans Hornung. (above)

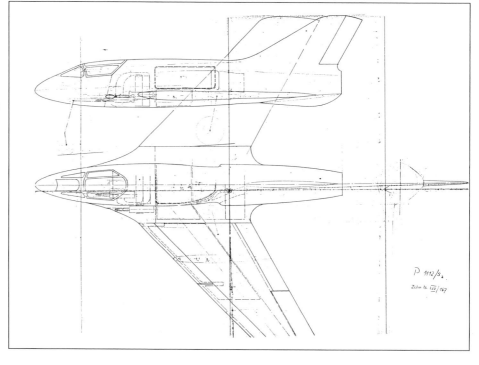

Incomplete P1112/S2 design: this undated photo shows lateral intakes. (below) P1112 S/1.

- The tail unit permits reduction of the wing surface from 22 m² to 19 m²

Further results and data can be gleaned from the designation and tables. Understandably, with the war's end approaching, it was only possible to transfer the designs graphically; the Oberammergau P1112 Bureaus could not draw from complete design documents, data logs, or performance calculations. It is improbable that the Messerschmitt documents or portions of his work on the P1112 with the RLM in Berlin were submitted or were even provided at the last conference for "A turbo-jet fighter" in March 1945 at Focke-Wulf in Bad Eilsen.

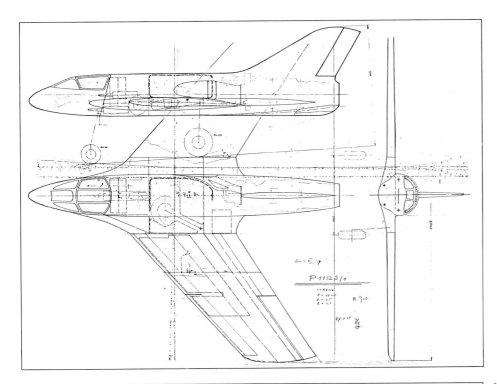

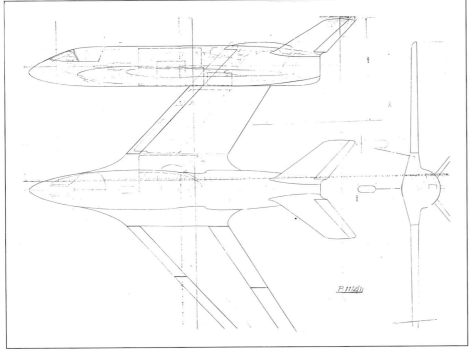

Hornung's continuing work on the P1112/S 2 study (above).

The final point: P1112-V1 from March 30, 1945

Abstract of the lines of development of the 1101/1110 and 1111/1112. (below)

The aircraft represented on the referred to full-view drawing XVIII/166 clearly shows the effort to connect the Messerschmitt led P1101/1110 and 1111 lines of development to date, and simultaneously, to avoid every risk (tailless configuration!) in reference to the guidelines from March, 1945,

With this design, the P1112-V1 designator apparently points to the fact that Woldemar Voigt wanted to reach the 1000 km/h barrier within one year, and yet "with certainty." Although the design could not get past its initial stage, many of its technical accomplishments are illustrated through the successful aircraft of the 1950s and 1960s. Neither a German design nor any type of foreign development of similar or equal specifications had obtained such status at this period in time.

This claim, daring in many respects, is emphasized by the following points:

- conventional concept with optimal, that is, resistance-free component configuration;
- configuration of the aircraft according to the area rule concept;
- highly swept wings with relatively low profile thickness and lift-increasing aids for verified low-speed flight; the low aspect ratio permits a stabile, anti-torsion wing construction;
- excellent, tested intake system using resistance-free platform intakes for engine air flow;
- large internal fuel capacity inside the fuselage and in self-sealing wing tanks;
- progressive, versatile weaponry.

Through consultation of a weapons study with 4 x Mk 108, completed on March 14, 1945, by Project Engineer Bächler under the designation P1110 W, work on the mock-up of the P1112-V1 forward fuselage section began in Building 607 of the Oberammergau complex. From a description of April 18, 1945, it read as follows:

"This mock-up represents the above named project (P1112) and is thought to be a provisional mock-up for configuration evaluations."

The fairing consists of egg-shaped curves and merges the junction to the fuselage center section with

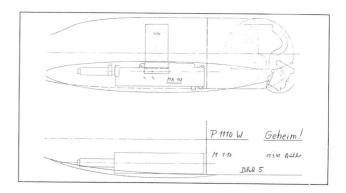

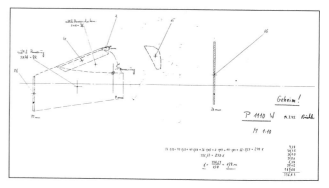

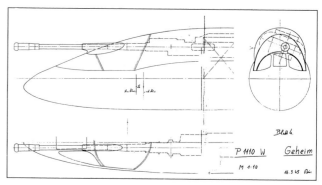

Weapons variant Mk 112 for the P1112. (above)

Amour of the P1112/P1110

circular cross-sections. Cockpit construction is not available and armored panels are set into the fuselage making forward vision sufficient and no substantial corners or inclines in the forward cockpit exist. The bullet-proof windshield is 100 mm thick and is inclined 23 degrees. The side, bullet-proof windshields have a thickness of 60 mm and are symmetrical, and equal for both the left and right sides. The shape of the remainder of the cockpit is developable and is subdivided with pressurized, rearview construction in connected sections.

For foot controls, a parallel control is to be used and the control unit is clearly mounted for excellent operability of both the left and right instrument panels.

The pilot seat is quite opposite from the normal configuration and is inclined to the rear, and nevertheless ensures tolerable seating.

Behind the pilot's seat 4 x Mk 108's are configured. Their protective tubes perform for the pilot as well as for the instrument panels.

Above and between the weapons are the ammunition containers, from which the ribbon fastenings lead straight to the weapons.

The equipment located behind the armament is easily accessible.

The nose wheel undercarriage lies behind the pilot and is normally retracted to the rear.

The war's end also consummated a portion of aviation history when, despite the pervasive shortages and the slightly increasing turmoil, highly qualified, creative aerodynamicists and designers plowed ahead into

Dummy construction for the P1112-V1 with 4 x Mk 108's.

From the end of March/beginning of April, 1945, in Oberammergau.

new, technical territory. And were thereby able to teach something to the rest of the technical world. Within the RLM's competitive bidding for a single-engine jet fighter aircraft, the Messerschmitt Firm continued their hard work during the months of July, 1944, until April, 1945, on almost completed prototypes, multiple fighter studies, and at least three progressive designs; with dummy construction of this aircraft beginning, continuation of the war, in the long outlook, would not have brought an equal match to the skies.

To emphasize, this was a gain for neither ideology nor the war, but the results of works from engineers and scientists who continually advanced knowledge and consistently put these ideas into practice. This purely technical work was continued, as is often mentioned, in the construction firms of the aviation industry of the victorious powers.

Geometrical data of the proposals:			
	P1111	P1112/S2 from March 3, 1945	P1112/S2 P1112/S1 from March 27, 1945
Wing span	9160 mm	7800 mm	8749 mm
Surface	28 m²	22 m²	22 m²
Aspect ratio	3.0	3.0	3.5
Relative thickness	8%	8%	8%
Sweep	45 degrees	45 degrees	40 degrees
Length	8920 mm	8200 mm	8250 mm
Height	3060 mm	3050 mm	3160 mm
Undercarriage			
Wheelbase	3130 mm		2100 mm
Tire size	740 x 210	740 x 210	740 x 210
	465 x 165	500 x 180	500 x 180
Armament	4 x Mk 108	4 x Mk 108	4 x Mk 108
Fuel	1200 kg	1550 kg	(1550 kg)

In the case of the P1112, the fundamental construction of the wing assembly is recognizable, above all, with the carrier-based aircraft of the U.S. Navy. For an example, the Chance Vought F7U-1 "Cutlass" can be named. This design, i.e. the "Cutlass" concept being based on the Arado documents, has been copied again and again.

As far as is known, the Arado projects concerned flying-wing studies with wing assemblies having wings swept to the outside, the crescent wing, or delta wing designs. The characteristic that reminds one of the Arado studies is the configuration of both vertical tail units on the wing assembly.

Final P1112 V-1 design for a standard fighter call for bids from the RLM, March, 30, 1945:

Crew	1 pilot in pressurized cockpit; protected and equipped with ejection seat
Engine thrust	1 x HeS 109-111 A-0 with 1300 kp static thrust
	or 1 x HeS 109-111 B-0 with 1500 kp static thrust
Length	8250 mm
Height	3160 mm
Wing span	8740 mm
Surface	22 m²
Aspect ratio	3.5
Relative thickness	8%
Wing sweep	40 degrees
Undercarriage	
Wheel base	2100 mm
Nose wheel	500 x 180
Main wheel	740 x 210
Armament	normal:
	4 x Mk 108 and approx. 500 kg external load
	reinforced weapons:
	1 x Mk 214 caliber 55 mm
	or 1 x Mk 112 caliber 55 mm with 70 rounds
Fuel	approx. 1900 liters, maximum: 2400 liters
Empty weight	2290 kg
Additional loading	2383 kg
Flying weight	4673 kg
External load	246 kg/m²

The swept wing of the "Cutlass" clearly shows the features of the P1112, and in the end, the director of the Oberammergau Project Bureau, Woldemar Voigt, found new employment at Chance Vought.

The Project Bureau could no longer draw up exact performance calculations until the end of the war.

An aircraft of this construction would surely have attained a flight speed of over 1000 km/h at an altitude of 7 kilometers, just as the P1112 would have matched the technical requirements of the "Special Commission Day Fighter" or the RLM's Technical Department.

The Competition

In the assessment following the conference from 27 to 28 February, 1945, a comparison between the single Messerschmitt proposal and the Junkers' EF 128 design. Although the Focke-Wulf project would be scantily mentioned in this report, Focke-Wulf evidently received a contract for prototype construction right along with Junkers and Messerschmitt.

After the war, a chronicler called the Focke-Wulf Ta 183 the "Successor to the Me 262." It can almost certainly be maintained that the RLM did not award an official production series contract in connection with this call for bids. The RLM number can hardly be an indication for award of a Ta 183 production series contract when one thinks of the Me 329, for example, from which is what Messerschmitt built a mock-up.

The operational capability of this construction type and the geographical set up of the manufacturing complex (Catchword: Fortress of the Alps) says much against favoring Focke-Wulf with a contract.

Much has been written about this topic and much of that has been contradictory; small wonder, when put into the perspective that, shortly before the collapse, the investigations were not simplified. What remained were the milestones which the designs put forth represented enroute to modern combat aircraft. Focke-Wulf set the lower wing load of the Ta 183 up against the higher speed of the Messerschmitt projects; through this they would attain higher altitudes and good low-speed flight characteristics. With the postwar programs of the MiG-15 (Focke-Wulf) and the North American F-86 "Sabre" (Messerschmitt), the Soviet Union and the USA carried on this design philosophy. The new type of aerodynamic construction of the EF 128 did not surface too often following the war. However, some known aircraft are the French flying-wing project SNCA Sud-Est SE-1800, substantially from Junkers' developmental director Professor Hertel, and, with limitations, the American carrier based aircraft Chance Vought F7U "Cutlass."

Both of the other submitted proposals from Blohm and Voss, and Heinkel apparently had no chance. The Henschel design was still not officially available in 1945, and, with its "risk afflicted" construction and a maximum speed of 984 km/h, it would hardly have had any prospects of overtaking the standard fighter of the Luftwaffe.

Two earlier American jet aircraft are also included the table. Both of these designs, which began their service in the Air Forces in 1945, were developed by the respective design teams (Clearance L. Johnson from Lockheed and Kendall Perkins from McDonnell) without knowledge of the aviation research completed in Germany. When comparing flight performance, the superior German aerodynamics are clearly noticeable.

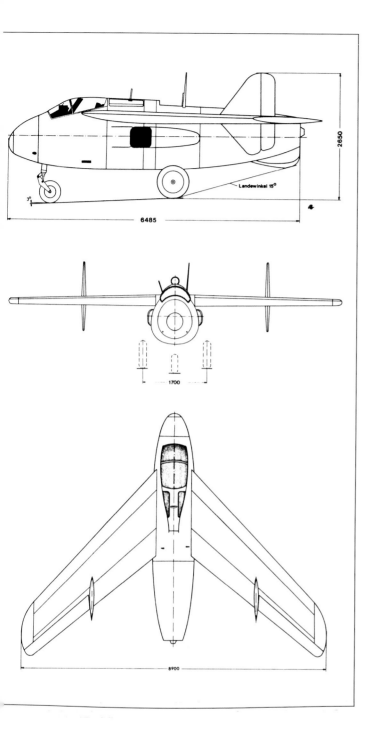

present while attempting to close in on the speed of sound were still unknown.

It can be noted that engine construction in Germany did not keep pace with airframe development due to the lack of materials (Nimonic!), while in the Anglo-American countries, the problem was the exact opposite.

Only the combination of both merits contributed to world success of the American aircraft industry after the Second World War.

Junkers EF-128: (According to Technical Report TB-Number 139/45, January 1945)

Specifications

In general: Tailless, high-wing monoplane with swept wings.

Wings: Twin-rail, all-wood wings, in two sections, with inner portion configured for taking on fuel.

Fuselage: All-metal fuselage. Nose fairing contains space for nose wheel undercarriage and additional, special equipment, connected to the pressurized cockpit as well as space for the main undercarriage and the air ducts. In the rear fuselage section lie the engine and another fuselage intake.

Tail unit: Aileron for horizontal rudder, the vertical tail on forward aileron section on both lower and upper edges.

Undercarriage: Nose undercarriage and main undercarriage as rocking lever undercarriage with flexible pneumatic shock absorber. Retracts pneumatically. Main wheel size: 710 x 185, Tail wheel for nose wheel, size: 465 x 165.

Engine: HeS 109-111 in fuselage rear. Accessibility ensured due to detachable portions of the fuselage surface. Intake openings for the intake air on side fuselage walls below the wings. Boundary layer separation provided. Intake of boundary layer air through layers at the end of the cockpit, fuel systems: unarmored in wings: 540 liters, the remaining 1030 liters in two armored tanks in fuselage.

Equipment: Equipped as standard fighter. Fighter-ejection seat, fire extinguisher.

Despite the higher engine power and the closely similar dimensions, the American designs fell decidedly short in the area of maximum speeds attained.

High-speed flight could not have been purely pleasurable, especially with the FH-1 "Phantom"; it was reported, "that a further increase in speed is absolutely impossible due to vibration, bucking, and instability", clearly indicative that the aerodynamic occurrences

Armament: 2 x Mk 108 with 100 rounds each in forward fuselage section, below the cockpit, additionally, possible use of 2 x Mk 108 with 100 rounds each.

Armor: Pilot protected against 12.7 mm rounds from front and 20 mm from the rear.

Junkers intended an elongated, two-seater configuration for the EF-128 for application as night- and bad-weather fighter.

Focke-Wulf Tank Ta 183: In all, three direct predecessors to the Ta 183 existed: Fw PV "Huckebein", FW PV Design 2 and Design 3, which, in comparisons to both of the other designs, is substantially more conventional. With Designs 2 and 3, Focke-Wulf took part in the RLM's call for bids and discussed the proposals at the conference from 12 to 15 January, 1945. The emphasis for this development is on Design 2, on items such as: design, data, and production documents

The Focke-Wulf Ta-183 V1; "Flying mock-up" with Jumo 004 B engine.

The intended aircraft is similar to the Messerschmitt P1101-V1 (Sketch: Sengfelder)

from February and March,1945. In March, when the construction contract for the prototypes was received, the jet fighter received the RLM designator Ta 183.

The technical specifications and data provided in the table refer to Design 2.

The prototype (V1) of Ta 183 was to carry out its maiden flight at approximately the same time as the P1101-V1, in May and June of 1945. For the first evaluations (V1-V3), the Junkers Jumo 004 B engine was provided, just as during the competition in Oberammergau, as the Heinkel turbo-jet engine HeS 011 was not available. And just as Professor Messerschmitt and Engineer Tank had planned, with the Ta 183 V1, an analysis of the different components was performed, such as the similar tail unit design of Design 3 (vertical tail II from March 29, 1945).

Immediately before the end of the war, the construction of the prototypes was as good as complete and the work complex began production of equipment and components.

The initial series, equipped with the Heinkel HeS 011 engine, would lead to the Ta 183 V4-V16 aircraft, whereas the V15 and V16 were intended for warehouses for statistical analysis.

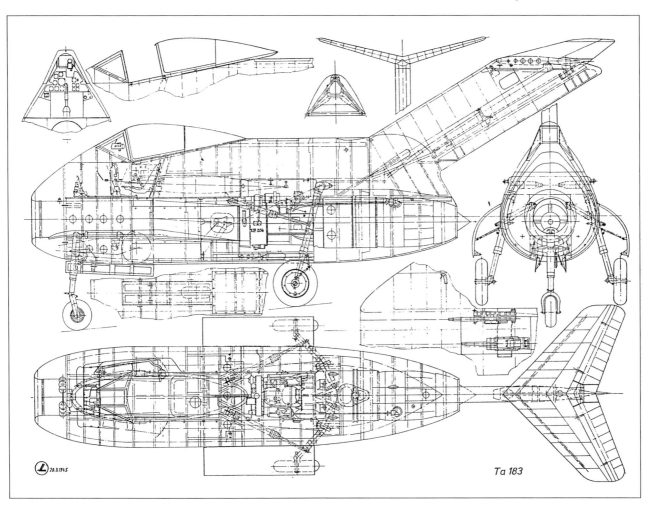

Ta 183

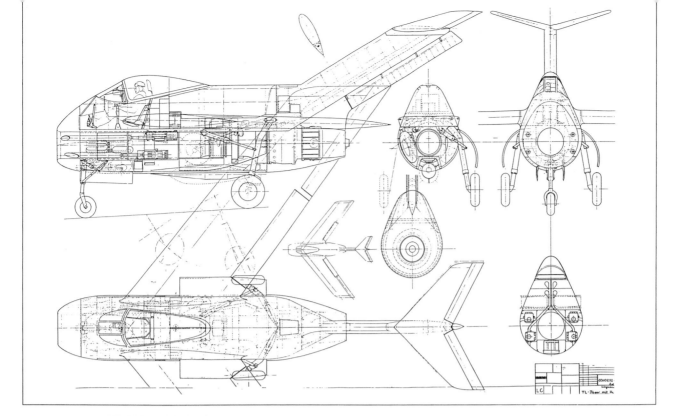

Focke-Wulf Ta-183, production series aircraft with HeS 011 engine. February, 1945 (above). Ta-183 model in wind tunnel (left).

The planners in Bad Eilsen set up the maiden flight of the Ta 183 V4, in a configuration from February 23, 1945, for the middle of August, 1945; the first Ta 183 A-0 production series aircraft were to roll out of their respective hangars in October of the same year, and a monthly output of 300 jet fighters was expected by July, 1946. This potential output would have been possible, according to specifications, with a total work force of 3,600.

Technical Specifications of Design 2 for Focke-Wulf Ta 183

(According to the layout plan from February 18, 1945 and the aircraft production specifications)

Primary function: Fighter aircraft for all altitudes, especially high altitude aircraft for flight at altitudes of up to 14 km.

In general: single-engine, cantilever, mid-wing monoplane with central, high-set tail unit and engine intake in fuselage nose.

Wings: Wing assembly without slat wings or leading edge flaps possesses a wing sweep of 40 degrees and a parallelogramatic form, it is constructed with steel spar booms and equipped with ailerons (lobe rudders) and landing flaps (trailing edge flaps). The wing assembly takes on part of the fuel; for this, either fuel density or its alternative, aluminum tank construction amounts to 0 degrees, the relative thickness 10%; the wing possesses neither aerodynamic nor geometric decalage.

Fuselage: The fuselage has an oval cross section with a circular faring, capable of arching to approximately 70% of the lower half. It is made of all-metal construction, partly steel (fuselage upper section, engine fairing), and partly duralumin. Near the pilot's pressurized cockpit, the engine, the military and electronic equipment, the fuselage also takes on the largest portion of fuel in either the fuselage's inclined middle section or the alternative, in the thinly walled fuel bladder tank.

Tail unit: Central vertical tail made of light metal with a relative thickness of 13% and a sweep of 60 degrees. The high set horizontal tail, constructed of wood, possesses a sweep of 40 degrees, just as does the wing assembly. The highly swept vertical tail, with a relatively high aspect ratio, would definitely have given the design problems during high-speed flight with the boundary layer flowing in the direction of the horizontal tail.

Undercarriage: Nose wheel undercarriage is a single strut mechanism with bent wheel fork, which is hydraulically retracted to the rear into the fuselage nose section; wheel size: 465 x 165 mm. The single strut undercarriage is retracted laterally into the fuse-

Dream concept for the Focke-Wulf developers: Ta-183 during operational testing.

Construction: G. Sengfelder.

lage; the shock absorbing strut was replaced by the Fw 190; wheel size 700 x 175 mm.

Engine: The engine lies inside the rear fuselage section and is easily accessible from the ground via a detachable engine cowling.

Fuel system: 1000 liters with sealed tank construction, or 850 liters with fuel tanks inside the fuselage center section; the remainder is stored unarmored inside the wings. Maximum fuel capacity possible: 2000 kg (approx. 2500 liters).

Equipment: Standard fighter aircraft equipment.

Armament: 4 x Mk 108 with 2 x 100 and 2 x 120 round configurations; ammunition is housed underneath the cockpit.

Armor: Cockpit protected from rounds of 12.7 mm from the front, and 20 mm from the rear.

Manufacturer Acft type	Messer-schmitt P1110	Messer-schmitt P1111	Junkers EF-128	Focke-Wulf Ta-183 Design 2	Lockheed P-80 A1	McDonnell FH-1 Phantom
Engine Static thrust	HeS 011 1300 kp	HeS 011 1300 kp	HeS 011 1300 kp	HeS011 1300 kp	J-33 A-9 1746 kp	J-30 WE 10 2 x 726 kp =1452 kp
Wing span	8.25 m	9.16 m	8.9 m	10.0 m	11.85 m	12.42 m
Sweep in t/4	40 degrees	45 degrees	45 degrees	40 degrees	-	-
Aspect ratio	4.29	3.0	4.5	4.45	6.36	6.06
Profile	0.08-40 0.12-40	0.08-40	0.10-45	0.10-40	65.2-013 65.2-010	
Wing surface	15.86 m²	28.03 m²	17.6 m²	22.5 m²	22.07 m²	25.45 m²
Total length	12.36 m	8.92 m	7.05 m	9.35 m	10.51 m	11.81 m
Height	3.18 m	3.06 m	2.05 m	3.48 m	3.45 m	4.32 m
Empty weight	2812 kg	2740 kg	2607 kg	2909 kg	3593 kg	3039 kg
Normal take-off weight	4290 kg	4282 kg	4077 kg	4379 kg	5307 kg	4524 kg
Normal fuel load, without additional tanks	1620 liters	1560 l	1560 l	1500 l	1609 l	1418 l
External stores during take-off	271kg/m²	153kg/m²	231.5kg/m²	195kg/m²	240.5kg/m²	178 kg/m²
Level flying speed	902 km/h at sea level 1000 km/h at 7km alt.	900 km/h sea level 995 km/h at 7km alt	905 km/h sea level 990 km/h at 7km alt.	875 km/h sea level 955 km/h at 7km alt.	898 km/h sea level 858 km/h at 6.1km alt.	771 km/h sea level 780 km/h at 4.57 km alt.
Initial climbing speed	21.5m./sec	2.37m/sec	22.9m/sec	20.5m/sec	21.2m/sec	21m/sec
Service approx. ceiling	14,000m	14,000m	13,750m	14,400m	13,716m	13,100m
Landing speed	178km/h	155km/h	186km/h	166km/h		
Armament	3 + 2 x Mk 108 additional jettinsonable weaponry	4 x Mk 108 additional 500 kg jettisonable weaponry	2 + 2 x Mk 108 additional jettisonable weaponry	2 + 2 x Mk 108 additional jettisonable weaponry	6 x 12.7 mm Colt Browning M2 and 454 kg jettisonable weaponry	4 x 12.7 mm Colt Browning
Operating range	1500 km	1500km	1300 km	1740 km max. with additional fuel	1770 km max. with additional fuel	1464 km max. with additional fuel
Remarks	main component test with P1101-V1	mock-up for further development started in Oberammergau	production series intended for mid-1945	Maiden flight planned Aug. 15, '45 production Oct. 15, '45	Maiden flight Jan. 8, '44 Delivery of P80-A beginning Jan. '45	Maiden flight Jan. 26, '45

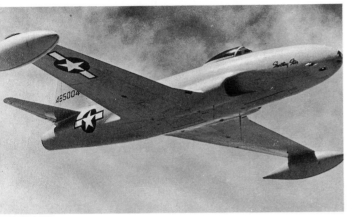

McDonnell FH-1 "Phantom" (above)

Lockheed F-80 A "Shooting Star" (below)

The Ta-183 served as the basis for the "Pulqui II", which Kurt Tank brought to life with his former Focke-Wulf team in Argentina. And earlier Soviet jet aircraft also show an undoubtedly strong resemblance to Tank's configuration.

In at least one area Engineer Tank appears to be won over by the quality of the Messerschmitt aerodynamics: the Pulqui II's wing design possessed similarities to the Ta 183, and contours were identical to those of the first wing set of the P1101 prototype.

Evaluations of the Designs

As stated, following the war, the proposals submitted to the RLM received extremely contradictory assessments. Messerschmitt's works surfaced with hardly an accurate appraisal. The technical data was often not even correct, and in order to complete the confusion, the various authors described "fantasy projects" such as the P1114 and the P1116. In place of a reliable assessment with facts and technical information, it was so terse that the RLM rejected the P1101 in the fall of 1944 and then chose the EF 128 and then the Ta-183 as the future production series aircraft. One

design was no more correct than the other.

The unabbreviated impression of the DVL statement from March, 1945, explains the actual conditions, and simultaneously permits insight into the technical and military problems with which they had to begin to deal with for jet propelled flight at subsonic speeds.

1. Performance

In terms of total performance ratings, Messerschmitt's P1111 project stands undoubtedly at the top. This design gave away the maximum speed rating to the P1110 (difference: 5km/h), while the P1111 was rated on top in all other calculated performance areas:

Climbing speed 23.7 m/sec as opposed to 23.2 m/sec with Focke-Wulf II and 21.5 m/sec with the P1110; take-off distance 600 m as opposed to 650 m for the Focke-Wulf II and 790 m for the P1110. Landing speed: 155 km/h as opposed to 164km/h for the Focke-Wulf I* and 178 km/h for the P1110. Landing distance 450 m as opposed to 490 m for Focke-Wulf II* and 610 m for the P1110.

* Authors' note: Focke-Wulf I was similar to the Ta-183 Design 2 and Focke-Wulf II was similar to the Ta-183 Design 3.

Junkers' design is evenly matched with the P1110 design from Messerschmitt: its maximum speed is approximately 10 km/h lower than the P1110 (rating 3rd for maximum speed), and the take-off and climb ratings are somewhat better and even lower than the P1110 in landing ratings.

In 4th place in the area of maximum speed, and 20 km/h behind the fastest, the Messerschmitt P1110 (under the assumption that the latter would loose 4% loss of thrust through the air intakes), lies Messerschmitt's P1101 with vmax=980 km/h. The take-off and climb performance ratings of these designs lie above those of the P1110 and are very near those of the Junkers' design, whose landing ratings are better than those of the P1110 and the EF 128.

The remaining designs are rated so far behind those named in speed ratings that they could be eliminated from the list of choices. The operational lessons learned up to this point showed that a performance superiority rating of about 30-50 km/h from the fastest possible aircraft would be an unendurable tactical disadvantage for a fighter aircraft.

The differences existing between the better designs' maximum speeds and, for example, the Focke-Wulf II design in regard to performance characteristics in take-off, climb, and landing, do not justify the difference in maximum speed. (Therefore, it is important to remember that with enlargement of a turbojet aircraft, the maximum speed increases equivalent to the thrust per m², which is very important concerning enemy bombers!).

With the relevance of the calculated performance ratings, the questions to ask is whether, with any of the prototypes named, the calculated performance ratings are tied to the considerable instability. The following should be noted:

a) Engine air intake

The greatest area of instability in the performance rating of all designs lies in the estimated loss of thrust at the engine air intakes of the designs of Junkers and Messerschmitt P1110. With both prototypes, the engine air intakes lie on the fuselage side rather far to the rear and direct the air to the engine at a considerable angle. With both designs, the inflow of the boundary layer in front of the engine should be neutralized by suction of this boundary layer, whereby Junkers would use suction by air flow pressure at an inclined area of the fuselage, and Messerschmitt, on the other hand, would use a special extractor fan. Messerschmitt's chosen configuration is more costly, however, it seems more reliable and is this probably associated with the lesser loss of thrust.

The significance of the performance instability in regard to boundary layer suction will be properly assessed, especially since the wind tunnel tests of low-speed flight do not supply corresponding data; it is certainly thinkable that boundary layer thickening, or separation of shock pressure before the engine intake would appear in high Mach numbers, and thus the ratio could be decidedly affected.

The loss of performance through the intake is calculated with 4% thrust. The P1110 design could endure a loss of 14 %, the EF128 9.5 %, until the speed sinks to that of the slowest of the four designs. This number is only valid for flight speeds at high altitudes (Mach influx) and only under the assumption that the Mach influx actually has an assumed sum in the comparison rating, and that the flight speed is sufficient to reach the steep drag increment range!

b) Deviations in the shape of the fuselage from the theoretical spindle shape.

The P1101 fuselage, with which the balance of performance on the basis of analyzing spindle shapes and circular cross-sections was rated, deviates considerably in realizing this shape. It is not considered ideal due, on one hand, to the profile because of the engine's configuration, and on the other hand, because of the cross section and its high frame; one must accept that these facts would somewhat unfavorably influence the performance compared to the rating.

c) The possibility of a slot-free wing.

Another drawback of the performance inquiry of the Junkers design is that this design is intended to be without slots (slotted nose flap instead). The slotted nose flap has still not, up to this point, been tested in actual flight, and otherwise it is to be used as an automatic prevention against flow separation. With the large wing shape of 45 degrees there exists a great possibility the this Junkers EF 128 design would need to be fitted with leading edge flaps as well, whereby the maximum speed would be pulled back approximately 10 km/h.

2. Flight Characteristics

a) Response to the lateral axis

The tail unit effect, in regards to stability and controllability, is considerably more favorable in the normal aircraft because of its larger level arm than in the tailless aircraft. This will allow for greater possibilities in equipment variation as well as more safety during unforeseen unbalancing of an aerodynamic or weight nature. Both of Messerschmitt's prototypes, the P1110 and the P1101, were superior to the EF 128 design from Junkers and Messerschmitt's own P1111.

With some designs, the moment of the loss of thrust at the center of gravity produces considerable displacement of the neutral point between full-throttle flight and idle. The Messerschmitt P1101 is one of the worst in this respect.

One critical standpoint is the fact that, with tailless aircraft, and in particular those with large aspect ratios, the effect of the elevator with ample impact pressure as a result of turning resilience of the wing can be severely reduced. Calculations, partly gathered during flight analysis, offer a reversal of the aileron effect for the 8-109 and 8-262 prototypes due to the wing's smoothness during speeds of approximately 1100 km/h near ground level.

This reduction, or reversal, of the aileron effect can represent a serious danger for tailless aircraft with large aspect ratios (EF 128). With the P1111 design, the danger can be considerably tempered with the low aspect ratio.

The response in high Mach numbers will be determined through a change in the center of pressure, and with the down draft changes at the area of the tail unit. The tailless aircraft are disadvantaged since their tail unit level arms are the same size as the profile thickness (which, for their part, is a dimension for the possible size of the change in the center of gravity). Aircraft with normal fuselage lengths and tail unit level arms would be less sensitive because of their greater stability and their better aileron effect. Actual flight testing was still lacking (the results of 8-163 are not to be seen as decisive, since they were carried out with thick and curved profiles).

The affect of the wing wake on the horizontal tail unit in high Mach numbers is low according to wind tunnel analysis on full-scale models and the follow-on measurements behind the wings for the remaining configurations. Flight tests by Messerschmitt up to Mach=0.86 resulted in no difficulties.

The P1101 design has a slight high-tail arrangement over the wing. This will certainly be raised.

b) Affect on the vertical axis and the fore and aft axis.

To assess the stalling characteristics of the aircraft under analysis, the following is offered:

	Junkers	P1101	P1110	P1111
Sweep	45°	40°	40°	45°
Î	10/10	8/12	8/12	8/8
1a/1i	0.57	0.524	0.524	0.3
Slot	-	Slot	Slot	Slot
Wing	17.6	15.8	15.8	28
b	8.9	8.25	8.25	9.16
l	4.5	4.29	4.29	3

A reliable assessment of the stall characteristics based on this table is impossible, since knowledge and experience with similar aircraft were lacking, especially for aircraft with larger wing sweeps. It could be said that the stall characteristics of wings swept to 40 degrees and more could cause certain risks.

The P1101 and P1110 designs lie on the safer side. These two designs have: the smallest wing sweeps; medium spiking; a thick outer profile; and a leading edge flap which extends extraordinarily deep onto the outer wing edge in order to avoid a localized stall at all costs. These advantages contrasted the other designs' low wing surface and wing span. Based on observations up to this point, it can be assumed that the leading edge flaps concealed these disadvantages, since their affect with highly swept wings plays a larger role than those selected.

The spin characteristics of the proposed prototype could hardly be foreseen. The lowest risk may be with the P1101, and the P1110 can be put at a disadvantage with these measurements.

There is hardly sufficient knowledge of the characteristics of tailless aircraft:

In summary, the following could be said :

The flight characteristic risks of the designs considered had little differences in stages of design.

The P1110 design is probably the safest of the aircraft: the spinning motion, which could be affected by in the longitudinal direction, has not been a decisive attribute for quite some time in assessing aircraft characteristics as a whole, at least according to our knowledge.

Somewhat more unfavorable is the P1101, in view of the low-level engine thrust, which allows approximately 13% neutral point displacement between full throttle and zero thrust.

Still somewhat riskier would be the Junkers EF 128 and the Messerschmitt P1111, which we also assessed. From the start, there is a high probability of the aircraft pitching due to the highly swept wings, and not from the leading edge flaps of the protected wings. And lastly, there are difficulties in consideration of the fact that there is very little knowledge and experience with tailless aircraft having extremely low aspect ratios (l = 3). For both of the tailless aircraft, there exists a danger of sensitivity around the lateral axis with regard to displacement of center of gravity of aerodynamic assessment and weakness of the wings with high dynamic pressure.

A test of the eliminated prototypes, assessed for performance reasons during further examinations, showed that none of them were superior, performance wise, and that acceptance of lower performance characteristics for a fighter aircraft was therefore justified.

3. Equipment

As stated from the start, the comparison of similar equipment was fundamental. This practice could not be used for all prototypes however, due to key problem areas in two cases where mightier armament was intended (or to avoid problems later when increasing the armament). Otherwise, there were differences with assessing the various designs due to the fact that the single prototype possessed, in part, different possibilities for increasing weaponry.

The number of weapons was normally configured as 2 x MK 108's. An increase in armament was intended as basic equipment for the P1111 (4 x MK 108) and P1110 (3 x MK 108) and is also calculated within the flight performance comparison.

The increaseability of the armament through installation of more weapons of the same configuration was not examined by the project designers. In this regard therefore, one must get by without an assessment.

An increase of armament through modification of the weapons configuration would make an assessment easier, as, for example, installing a heavy MK 112 in place of several lighter weapons. In this regard, the Junkers EF 128 and Messerschmitt P1110 are the most favorable. These configurations allow a wide permissibility of up to approximately 350 kg or 400 kg center of gravity housing capability in the sealed weapons storage areas. The P1111 is somewhat worse off, since it possesses a sealed weapons area and is fundamentally more restricted in the realm of center of gravity workload utilization (the lower configuration weight of the turret and the tailless construction).

The P1101, with which the housing of 4 x MK 108's is possible, is the least favorable, with its configuration allowing no fusion of the weights for larger single weapons.

Housing for the search device, or seeker, and launch devices is by far most possible with the Junkers design.

4. Engine System

Junker's engine configurations, specifically the P1110 and P1111, have a great advantage over the other designs, and especially the P1110, because the engines are well protected from being fired upon from the front due to their configuration in the center of the fuselage rear. And, they're also easily maintainable.

Several of the submitted designs have fuel tanks in the wings that cannot be dismantled:

P1111	100% fuel in undetectable wing fuel tanks
EF 128	35%
P1101	25%

The wing fuel tanks are somewhat at a disadvantage for removal due to their thin and smaller wing sections (large, flat bottom to ensure relatively large residual quantity, unfavorable installation of instruments). With tailless aircraft, in addition, problems also occur since the center-of-gravity sensitivity of this construction type demands distinct precautionary measures that ensure an even, center-of-gravity extraction from all fuel tanks and prevents a rolling motion of the fuel.

The area for a fuel capacity increase was only foreseen for the P1110; the wing can hold approximately 400 liters.

Protection of the fuel systems is varied. One must take the given thin wing thicknesses into account, since the entire fuel capacity inside the wings is unprotected; this standpoint is the greatest significance for the assessment of all of the aircraft's data. The best aircraft to assess would be the P1110, where the entire fuel capacity is protected, and in addition, the fuel tank lies inside the fuselage, behind the cockpit and weapons storage area and the engine. The second safest would be the P1101 design, where 75% of the fuel is located in the fuselage and is concealed by the armored cockpit and therefore, protected from the front.

The remaining 25% is located in the wings and unprotected, however with the engine and the cockpit covering the armored fuel tank, it is well housed.

Design EF 128 has 35% of its fuel in unprotected tanks and another portion in the tank installed around the engine (It is recommend to omit this fuel tank).

In this regard the P1111 design is the most unfavorable, since the entire fuel capacity in the wings is unprotected so much so that it offers the enemy a target area of 0.85 m2 from the front and 17 m2 from the rear. Militarily speaking, we do not consider this configuration acceptable.

The engine intakes of the four comparable prototypes are assessed as follows:

The P1101 intake is simple and without any risks though the P1111 is somewhat more complicated, and may not have any large risks attached, according to analysis by Focke-Wulf.

The intake configurations of the EF 128 and P1110 designs are, in this respect, extremely risky, since with the imperfect resolutions of the suction configurations, considerable engine output loss can occur; in section 1) the output being cited would consist of significant reduction in engine performance being acceptable, but only if it doesn't endanger the value of the prototype. In this respect, the Junkers configuration appears less safe, and the system chosen by Messerschmitt calls for increased expenditure in the form of subsequently required suction fans.

Summary

A summarized assessment of the engine systems is simply not possible, since the best total configuration with technical soundness is, unfortunately, also burdened with considerable technical risks.

During the assessment of the total configuration of engine and fuel systems, the P1110 design appears to be the most favorable;

The engine is protected from the front and installed in the fuselage rear. The entire fuel tank system consists of armor protected tanks which are easily detachable, configured in the fuselage, and also concealed and protected from the front and rear by the armored cockpit and the engine. On the other hand, the P1110, as stated above, carried the greatest risks in technical failures (together with the EF 128 from the configuration of the intake openings). And in addition, the expenditure from the configuration of the suction fans, driven from the engine, is more than with the other aircraft.

In case the P1110 is ruled out because of risks from the engine intakes, the best total configuration evaluated is the P1101 proposal:

Protection of the engine and fuel tank systems is much lower then with the already reviewed project. This design also had a decisive advantage, as opposed to the other designs, with the completely risk-free engine system and the possibility of protecting the entire fuel capacity (the wing tanks, with 25% of the total fuel capacity, are situated in such a manner that their draining can be accomplished before beginning a dog fight). In addition, the spacious tank is sheltered quite well in the center of the fuselage.

The P1111 design was eliminated as a possibility due to its not being able to protect the fuel.

5. Other Features

The pilot's view is, with all configurations, quite the same.

The possibility of the pilot ejection is best suited by the EF 128, since, behind the cockpit, there is no vertical tail. The other designs may be somewhat similar in this respect.

Conclusion

The total assessment of the singular designs provides the following:

1. The three designs with adequate maximum speeds, P1110, P1111, and EF 128, are hampered with risky arrangements (engine intake) and drawbacks (unprotected fuel tanks), which did not permit their being favorably considered.

2. The remaining designs are so strongly inferior in maximum speed that a final judgment in their favor appears impossible, since such a judgment would constitute the operability of the aircraft attaining the technical superiority (over the enemy, too!).

3. The performance results of the P1111 design seem to bring light to the fact that, in comparison to the normal aircraft and the tailless aircraft with a relatively large fuselage, a substantially better combination would be of maximum speed and take-off, climbing, and landing characteristics by developing a flying-wing aircraft.

The following judgment is proposed:

1. Placing an order with the immediate goal of establishing a production series can still not be granted.

2. The P1101 study aircraft, presently under construction, is to be completed and tested as quickly as possible; the same is valid for other study models under construction.

3. New designs are to be drawn up for the single turbo-jet engine obtained in the production series. This would merge the performances of the, up to this point, best design, and do away with risks or obstacles.

A proposal for the technical requirements is provided:

4. Based on the designs provided up to this point,

only three firms received the project order; with this limitation, a fragmentation of the design capacity should be avoided by overlapping work of the individual firms, as with a concentration of the German development on the tasking for a single turbo-jet engine in favor of another military tasking or other, more important tasking.*

* Author's note: In the Main Development Commission (EHK) they obviously have other ideas: Focke-Wulf Project Number 279 (TA-183, Design 2) was considered an immediate solution (which was rejected by TLR/F1-E) and Messerschmitt was to bring along an "optimal solution" at a later point in time. Somewhat later, in the first weeks of March, TLR/F1-E called for nothing less than the development of five projects. (Junkers, Blohm and Voss, Focke-Wulf, Messerschmitt and Henschel/Lippisch.)

In comparison to the other presentations from the middle and end of 1944, the Technical Bureau of the RLM expanded and modified the configurations once more for future jet fighters at the beginning of 1945.

The proposal mentioned in the "Conclusion" of the "Special Commission Day Fighter" corresponds to the following:

Technical Guidelines
(Proposal of the Special Commission Day Fighter)

1. In general:

Equipment: The standard fighter equipment is: control of heading, pressurized cabin, EZ 42.

Weaponry: 2, but up to 4 at the most, x MK 108 with 100 rounds each, or 2 to 4 MG 213/30 with 150 rounds each. Desired is the possibility of installing increased weaponry, like the MK 103, MK 112, etc. Additional weaponry, if possible, is intended.

Release load: For excessive load, the inclusion of 500 kg exterior load is called for.

Armor: The pilot and the MK 108 ammo area are protected from shelling from the front against 12.7 mm and from the rear against 20 mm. Sound protection for the engine against shelling from the front is to be attempted.

Fuel: Normal fuel capacity is at least 1200 kg = approximately 1500 liters. To be tested is how far the configurations of the General Staff and the total fuel capacity of the aircraft are sustainable for 2 hours flight time with 100% thrust at 9km altitude. 2/3 of the fuel weight is well protected.

Communications System: According to the Chief, TLR.

Increaseability: It would be desirable that a later enlargement of the wings and the payload volume would be possible, through larger fuel capacities, expanded equipment, or a more powerful engine (up to 2000 kg thrust).

Performance: The boundaries for the performance calculations depend on the single jet-engine performance comparison, which is the calculation method used by the DVL. This corresponds to the respective status of the research and development, and is matched with the knowledge gained from flight testing.

The following performance ratings are desired (with normal fuel weight):

Horizontal
Speed: At its best speed at an altitude not under 7 km, and with half the fuel weight under consideration of the Mach effect with 120% thrust, the minimum should be: vmax greater than or equal to 1050 km/hour at sea level with 100 % thrust v max greater than or equal to 900 km/hour.

Climbing speed
at sea level: With take-off weight minus fuel for engine warm up, take-off, and acceleration, with 100% thrust, it should be: w > 20 meters/second

Take-off: The take-off distance, without overloading, with 100% thrust should not be over 650 m ($\hat{E}$ = 0.04).

Landing: Landing weight with one-third fuel weight and total munitions weight: Landing speed should be as far under 165 km/h as is possible.

Landing run distance: With wheels braked, should be as short as 555 m as possible.

Equipment cost and risks: The material given by Director, TLR, or EHK prior to construction start will be committed. Attempt to bring design from its initial stage, as quickly as possible, to full development and at the lowest possible production cost. Designs possessing substantial risks of any sort are to be avoided.

These prerequisites formed the basis for the final Ta 183 design and for Messerschmitt's P1112.

Approximately ten years later, NATO issued specifications for a light fighter which emerged, in large part, from the knowledge gained during the Korean War, and concerned aircraft put into action in Europe.

The superior areas led to this call for bids and were included in the prerequisite data, even if in a totally different situation, not unsimilar from 1944: NATO wanted to counteract the continuously rising costs and the escalating take-off weights. The key would be a less costly, uncomplicated, and easily maintainable aircraft. The technical department under General Norstad at SHAPE (Supreme Headquarters of the Allied Powers in Europe, Belgium—NATO Headquarters) called for a single seat aircraft with a pressurized cockpit and internal weaponry of 4 x 0.5" MG or 2 x 20 mm Mk or 2 x 30 mm Mk and diverse external load available (3" rockets). The empty weight should not exceed 2250 kg and take-off should be possible over a 15 m obstacle within 1000 m on a grass runway. At ground level, SHAPE desired near Mach speed (Mach 0.95); and further demanded performance features were an average cruising speed of 640 km/h and a taxi speed of 100 degrees/second at Mach 0.9.

West European firms developed a total of nine proposals with these specifications. Five competitors came from France, among which, the Breguet Br. 1001 "Taon" and the MD "Etendard VI"; Great Britain brought SHAPE two designs, one being the Folland FO. 139 "Midge" (later called the "Gnat") and Italy sent the Aerfer "Sagittario", and last but not least, the Fiat G. 91 into the running. In October, 1957, NATO announced the Fiat G. 91, which completed its maiden flight on September 8, 1956, as the victor for the future production series aircraft.

Almost four years passed by since the public announcement of the call for bids in a newly freed Europe.

The comment should be allowed that the German aviation industry, around the time of the end of the war, had only nine short months for development of a similar type of aircraft. Certainly Germany still required some time until the appearance of the P1112 or the Ta 183, and to be sure, the predecessors, developed within the RLM or NATO, are only partly comparable with one another. And competition yielded some information of the achievements of German aircraft manufacturers in the years of 1944 and 1945.

A Breakthrough, but not an End

Twilight of the Gods in Oberammergau

Due to the swift advancement of the allied forces, those responsible at the Oberammergau Research Complex ordered the transfer of the plant to Kalkül; they wanted to eventually move further south, to Tirol.

Many important documents were recorded on microfilm and, together with further documentation, calculations, and designs, totaling 19 jet exhaust pipes and 45 fuel tanks, they were packed in watertight containers. Also originating about this time frame was the last technical document available to the authors from Oberammergau: a description of the mock-up of the fuselage front end of the P1112 from April 18, 1945. Prior to the invasion by the Americans, many of the Messerschmitt workers dispersed to the surrounding villages. The containers with the secret documents were put into four hiding places.

On a Sunday, it is written that on the 29th of April, 1945, an American infantry unit reached Oberammergau and took hold of the documents from the Messerschmitt Firm, without immediately recognizing the significance. This correlates to the fact that allied aerial reconnaissance never detected the Oberammergau complex and therefore never identified it either. Following this seizure, the GIs properly plundered the buildings and hangars, supposedly to create space for an "I and E School"; sometimes the space was also used for billeting.

An eyewitness describes: "They rummaged through all desks and cabinets, took what seemed worthwhile to them, and smashed all the supplies to smithereens with axes, turning them into tiny pieces."

During these actions, the soldiers also pulled the P1101 experimental aircraft from its tunnel.

The damages, clearly visible on the fuselage nose, came from either the obviously non-expert treatment from the military (from the axes?), or from a few of

Newspaper report from 1945: the picture shows the 1101-V1 with an inscription mounted by the Americans, having been pulled from its tunnel. In the background is the engineless Me-262, which was located in a protective bunker in Oberammergau awaiting conversion.

the Messerschmitt workers trying to destroy the prototype shortly before the American invasion.

The damaged birds were photographed by a war correspondent with a "conqueror."

The picture came with the headline "Mysterious Nazi Aircraft Found in Oberammergau Mountain Plant" in the American press.

With these aircraft, precarious even for the American technicians, the victors wanted to show, to the amazed population, the progressive lunacy of the Germans; the actual meaning wasn't immediately known. In the photo's caption, the aircraft was designated as "Mock-up made out of wood an aluminum", this was a false estimation which is sometimes still repeated in

technicians. They came to Oberammergau on the 7th of May, 1945, at least a week after the military's seizure of the complex. A delegation from the CAFT (Combined Advance Field Team) was first supposed to estimate the significance of the complex and classify it accordingly.

On the 17th of May, 1945, another group of experts arrived in the Bavarian mountain village, these from the USSTAF-ATI (United States Strategic Air Forces-Air Technical Intelligence). And a few days later, on May 21, 1945, a man met with Field Team 85 who was to take on a leading roll in the registering and investigating of the Oberammergau complex; Robert J. Woods, Technical Director and Co-Founder

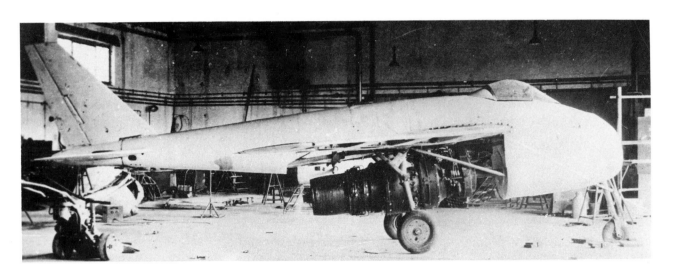

today's literature.

The jet fighter was once again brought into Hangar Number 615, where it was scoured by American

of Bell Aircraft Works, and later editor of an all-encompassing report about the Messerschmitt Factory (the so-called Woods Report). He was also the initiator of the Bell high-speed and swept-wing research. With Woods, two Bell employees were also involved in Oberammergau: R. Kluge and H.V. Hawkins; in the meantime, the representatives of the noted English and American aircraft manufacturers and equipment firms came into the picture to evaluate the documents.

In a hotel (Haus Osterbichl), the Americans detained key members of the Messerschmitt Firm whom they were able to spot, with others still remaining in the village and some in the surrounding area of Oberammergau. The "interrogations", that is, the thorough questionings, took place mostly in this hotel.

It is totally understandable that the first and foremost efforts were to uncover the hidden documents. But with the first and most complete hiding place in the cellar of a house in Wertach/Allgäu, there was a nasty surprise for the Americans: units of the First French Army had pushed their way into Allgäu and occupied the village Wertach. The uncovering of the hiding place was quite simple for the French, after it was explained to them from where to raise the treasure. The 23 metal containers found and the 8 steel tubes with layouts were promptly sent to the Ministry of Aviation in Paris. The materials captured were surely of great help for the French Aviation Industry, which, immediately following the war, endeavored to join the international aviation world. Only sometime later were the captured documents made available to the Americans. And among the documents were manufacturing data for the P1101-V1.

Two other hiding places, in the surrounding area of Oberammergau, were excavated by the Americans,

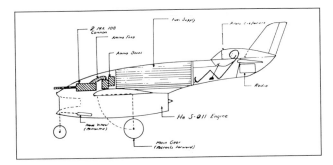

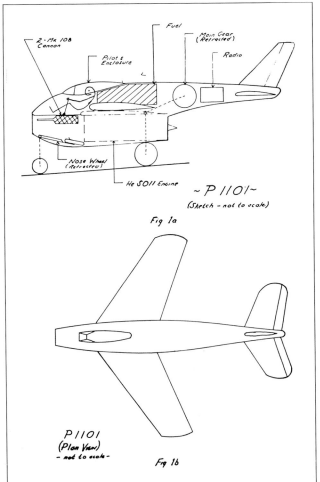

while hiding place number 4 was tracked down by an English captain.

Everything turned up with the four metal containers.

Help Yourself

The results of the extensive questioning of the key players from Messerschmitt by the American and British technicians was summarized by the civilian leader of the commission, Robert J. Woods, in his already mentioned, all-encompassing report (Review Number F-IR-6-RE).

Furthermore, the Project Bureau's forward-thinking fighter designs, in a more or less roughly sketched form, turned up in a report by the U.S. Navy which the CIOS in London disseminated to interested entities. With this report, designated CNO (OP-16-PT), one could acquire, at best, a rough overview.

Bit by bit the results of the German work found their way to the remaining aviation-driven nations.

Even in neutral Switzerland, not too far from Oberammergau, they were busy in the post-war period with the evaluation of German research reports. From where did the Swiss get these documents? That

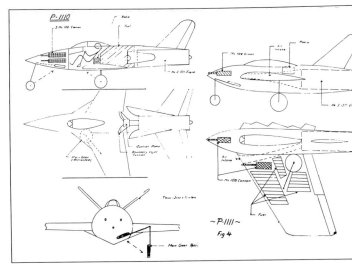

Sketches from the CIOS Report CNO (OP - 16 - PT) from the U.S. Department of the Navy.

could be speculated: immediately before the end of the war, the SS had detained a Messerschmitt employee who had attempted to smuggle documents into Switzerland. Only the end of the war saved him from execution. Whether or not this failed ordeal was the

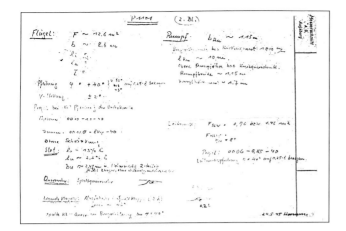

first attempt, or if the employee of that sort of "errand" had been successful before could not be explained.

In the fall of 1945, the Project Director of Swedish SAAB Aircraft Works, Frid Wanström arrived in Switzerland and evaluated the German documents there. Wanström immediately recognized the signifi-

Inspired by the P1111:

Experimentation Aircraft

SAAB-210, a predecessor of the SAAB J-35 "Draken" and SAAB J-35 (below).

Two examples from the works of Hans Hornung in the summer of 1945 under American direction:

- Specification sheet from the P1101: "Commemorative Report" from May 24, 1945 (above).

- Flight performance characteristics of the P1101: Explanation of the swept-wing effect from June 16, 1945 (below).

cance of the contents in front of him, and the Swede successfully translated the contents. Some of the former Messerschmitt employees made their way to Sweden to SAAB Works in Linköping (like Stress Analyst Dr. Behrbohm, for example). Under closer scrutiny, with the successful construction of the SAAB J-29 "Tunan", SAAB J-32 "Lansen" and SAAB J-35 "Draken" (and SAAB 210), several similarities with the Messerschmitt P1101, P1110, and P1111/1112 Projects can be established.

What is interesting in this summary is that a letter to the editor, printed in the British magazine "Air Pictorial" in August, 1966, stated therein, among other topics that: "In his article about the Tu-104, Boffin writes that the SAAB J-29 was connected with the Focke-Wulf Project (meaning the Ta-183, Design 3). During a short visit to the SAAB Works, I was informed that the J-29 originated from the Me P1101. SAAB developed the concept further and even improved it substantially."

Robert J. Woods Continues the Work:
From the P1101 to the Bell X-5

Sometime around the end of May and the beginning of June, 1945, the work on the experimental aircraft was restarted. R. J. Woods was of the same mind as W. Voigt: specifically that the aircraft should be repaired, completed, and returned to flying status.

These plans proved to be much more difficult than expected since, among other things, many of the design documents were destroyed or transported to Paris, and the French continuously refused, at this point in time, to hand over the documents to the Americans.

In short, the aircraft was never completed, and with its status of: having a battered fuselage nose; not being equipped with an engine cowling nor cover for the inner wing. And saddled with a more than likely unflyable Heinkel engine (probably a mock-up engine), the disassembled aircraft, packed in crates, nevertheless made its way via rail and ship to the United States. To be more exact, to Wright Air Force Base in Dayton, Ohio.

The new owners, the United States Air Force, stored the aircraft. In the coming months and years, Messerschmitt's ambitious project was to become an on-again, off-again object of intense study.

With this particular exploitation of technical aviation from the spoils of war, Wright AFB had seized an object whose worth probably still cannot be estimated. In the first few months following the war, many of the leading personalities of the crushed German aviation industry arrived at Wright, mostly via Wimbledon, England: therefore, among others, Woldemar Voigt, Alexander Lippisch, and Richard Vogt, i.e. the scientists and engineers in Germany who had worked on realizing Busemann's ideas and who had not hidden the disadvantages of the new type of wing; the swept wing. They and many others worked until war's end, and with success, to improve the low-speed flight characteristics of the new wing assembly. As is often the case, many promising possibilities were also recom-

mended here:

- Equipping the swept wing with lifting aids, such as leading edge flaps, slots, landing flaps, etc.;
- Special positions for the swept wing, like forward sweeping or delta-shaped wing;
- and last but not least, the wing assembly with various sweep settings.

All of these trials were tested in the various research and testing institutions and were tested on certain, comparable aircraft.

The German "Developmental Assistants" came to the United States in the framework of "Operation Paperclip" and spent the first months and even years writing reports of their earlier works in Germany and occasionally assessing the newest American designs. Most of them sought follow-on work in American industry and for many, the USA became their new homeland.

Back in Germany, however, the total capitulation of the aviation industry had occurred. Though the German influence of forward-looking technology on the leading industrial nations is still even today incalculable.

In the American aircraft industry, partly under pressure from the military, the following was swiftly proceeding as well:

- with North American, the crude jet fighter FJ-1/ XP-86, developed from the P-51 "Mustang", was converted into the magnificent F-86 "Sabre", propelled into existence with the aid of the Messerschmitt documents;
- and similarly with Republic: the F-84 "Thunderjet", derived from the P-47 "Thunderbolt", was further developed into the F-84 F "Thunderstreak" following divulgement of the successful German outcome with swept wings;
- McDonnell converted the design of a long-range fighter to one with swept wings and the prototype received the designation XF-88: the basic design for the F-101 had emerged;
- with Douglas and Convair, combat aircraft appeared with a new type of wing assembly; the delta wing from Dr. Alexander Lippisch;
- Chance Vought developed a design without a horizontal tail but with swept wings, the F7U "Cutlass", just as they had seen them in the German sketches and design documents;

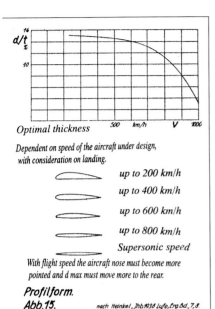

Optimal thickness

Dependent on speed of the aircraft under design, with consideration on landing.

up to 200 km/h
up to 400 km/h
up to 600 km/h
up to 800 km/h
Supersonic speed

With flight speed the aircraft nose must become more pointed and d max must move more to the rear.

Profilform.
Abb.15. nach Heinkel, Jhb.1938 Lufo.Erg.Bd. 7, 8.
Profilauswahl abh.v.d.Geschwindigkt.

Key to Diagram:
Profilform= profile
Abb. = illustration
Profile selection dependent on speed According to Heinkel, 1938, Aviation Research Results, Volumes 7, 8.

Great Britain, there was a limit on the thoughts of high-speed flight: the two aerodynamic measures that would lead to reaching the goal of Mach 1 should be: the aircraft should receive a projectile shape and a very thin, straight, wing with a symmetrical profile.

With these measures and with a powerful rocket engine, the Bell X-1 smashed through the sonic barrier on October 14, 1947.

At Miles in England, the designers attempted to follow suit in similar manner with the Miles M.52, which would have led to reaching its goal with consistent work.

In the Anglo-American countries there was nothing definitive in regards to realization of the swept wing. Shortly before war's end, aerodynamicist Robert Jones had proposed a swept wing assembly to the NACA for high-speed aircraft; he was snubbed for pursuing "wild ideas"; apparently his empirical works had not undergone any scientific verification.

The extent to which Jones was familiar with Busemann's theory is open to question; after all, the Volta Congress of 1935 had not been a closed-door conference.

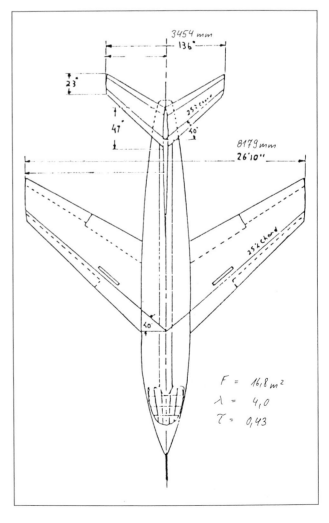

- At Boeing, Project 432 made its transformation from Model 448 to Model 450-1-1, later to become the Boeing B-47. Both the 448 and 450-1-1 designs already possessed swept wings, which Boeing engineer Shairer had retrieved from Braunschweig-Völkenrode in 1945.

- and not to be forgotten in this incomplete recapitulation is the immense contribution achieved by German aerodynamic research on high-speed programs for the U.S. Air Force and the U.S. Navy. Messerschmitt's specific contributions will be delved into more deeply later.

In the USA, the high-speed research had also already begun by the end of the 1930s. But just as in

The initial works for much research obviously came from Germany. American scientists undertook the sweeping concepts of their German colleagues and carried out their work in practical evaluations in such a way that the German research scientists could hardly recognize it.

A result of this work was the proposal of aerodynamicists immediately following the war for the planned experimental aircraft Bell X-1 to undergo a series of modifications. And these modifications were corroborated by tests in the wind tunnel:

- V-shaped tail unit;
- Wing assembly with both positive and negative sweep settings;
- Tiltable wings for the Messerschmitt-proposed combat aircraft construction type, which NACA scientist Charles Donlan analyzed in the Langley Research Center in a 7' by 10' (2.13 m x 3.048 m) wind tunnel.

V-shaped tail unit proposal for the Bell X-1; Messerschmitt had preferred this tail unit design for his high-speed designs during the war. (left)

Bell X-2, triple the speed of sound with its wing assembly shape similar to Messerschmitt's "Wing Assembly A." (below)

In October of 1945, in place of a similarly modified Bell X-1 aircraft, the proposal was made for a totally new swept-wing aircraft, and on December 14, 1945, Bell received a development contract for two "Supersonic Swept-Wing Research Aircraft XS-2."

Charles Donlan, just as with the Bell X-1, proposed that the new research aircraft X-2 have its tiltable wings tested by him in the Langley Research Center. The NACA was still not interested.

In the summer of 1946, the Bell X-2, with its intended swept wing, was tested with two similarly equipped Bell P-63 A fighter aircraft. These aircraft carried the designation L39-1 (Number 90060) and L39-1 (Number 90061). The flight tests carried out proved the usefulness and workability of the concept and formed the basis for wing construction of the Navy's counterpart to the Bell X-2: the Douglas D-558-2 "Skyrocket."

The Bell X-2, outfitted with a wing assembly similar to the Messerschmitt "Wing Assembly A" (similar in its conspicuous geometric shape), was awarded with an official construction order on July 3, 1947: however, the test program did not begin until June 27, 1952, with the first gliding flight, at a time when the Bell X-5 had long been flown.

On May 12, 1953, the X-2 program suffered a major setback when the first Bell X-2 (46-675) exploded on its B-50 launching aircraft and crashed into the Lake Ontario from an altitude of over 9000 meters. Test pilot Jean Ziegler, who had also flown the Bell

X-5 for its maiden flight, perished in the crash.

The first "real" flight, that is to say, with the rocket engine, only took place on November 18, 1955, and one year later, on September 27, 1956, the final flight of the Bell X-2 took place, as it had become difficult to fly and especially to land. During this flight, however, an unofficial work speed record was reached: Mach 3.2; the flight ended with the death of test pilot Milburn Apt. This speed was not to be outperformed until 1961.

Technical Data of the Bell X-2

Primary function:	single-seat research aircraft (swept wing, high-speed), the main point being its con struction of rust-free steel and metal alloy "K-Monel"
Length:	13,840 mm
Height:	3580 mm
Wing span:	9830 mm
Wing surface:	24.15 m^2
Profile:	from the NACA designed, bi- convex airfoil circular profile with the designation 2S - (50)(05) - (50)(05); that is to say, the wing maintains a rela tive thickness of 7.66% in 50% wing chord.
Aspect ratio:	= 4
Sweep:	40 degrees
(t/4-line)	Due to the flutter stability, the wing possesses a reinforced aileron, and for increase in lift during low-speed flight, a lead ing edge flap on the outer wing edge; both were originally Messerschmitt ideas.
Engine:	A twin-chamber rocket engine, Curtiss-Wright XL R25-CW3 with a thrust regulated between 1200 kp and 6800 kp.
Empty weight:	5613 kg
Take-off weight:	11,300 kg
Flight characteristics:	Maximum speed: 3370 km/h (Mach 3.2) Maximum altitude: 38,450 meters Flight time with full-throttle approximately 11 minutes

With construction of the Bell X-2 being halted, the tiltable wing proposal was never moved ahead. The wind tunnel experiments, conducted by NACA at Langley and already in progress for three years, yielded no practical conversion. In this situation, dur ing the summer of 1948, Bell's Chief Engineer Woods was convinced of the concept of the variable geomet ric wing and proposed an enlarged and modified de sign to the Messerschmitt P1101, as the aircraft should have been able to have the sweep angle of the wing assembly modified during flight.

Consequently, as was later proven, adjustment of the wing sweep during flight was to bring the results proposed by Messerschmitt. This concept concerned a separated wing with which, while being rotated, the center-of-gravity displacement from a back-and-forth wing movement of the entire wing assembly would be compensated for. Woods intended to further use the original P1101 as a flying test bed for a new gen eration of American jet engines. Not only turbo, air- breathing jets, as during the 1950s with the mass pro- duced General Electric J-47, or the somewhat unlucky Westinghouse constructions J-34 and J-46, but the later Allison J-35, installed in the X-5, was to participate with the experimental aircraft form Oberammergau in a flight test. In addition, those running the program had even planned to convert the P1101 into a ram-jet driven aircraft; they retooled it (with a cowling on the nose intake) into a pure rocket-propelled aircraft, as they had planned to move ahead.

However: this all remained only theory. During the hand over from Wright AFB to Bell Aircraft Works in Buffalo, N.Y., in August, 1948, the German aircraft fell off a freight car. This led to considerable damages which made a flight test totally impossible. Between this and technical problems uncovered during a more in-depth probe of installation and structural alterations, all plans had to be scrapped.

The damaged bird remained grounded and Bell technicians used the construction for adaptive work with the X-5 engine — the already mentioned Allison J-35. They also fastened mock-up weapons on the side of the fuselage (4 x MK 108 and, incomprehensibly, 6 x MG 151).
J-35. They also fastened mock-up weapons on the side of the fuselage (4 x MK 108 and, incomprehensibly, 6 x MG 151).

Following unfavorable evaluations from experts, the U.S. Air Force declined utilization of the fighter interceptor with variable geometric wings for military purposes (Intended quantity: 24). Though the Air Force, just as the NACA, had originally approved construction of the X-5.

The principal objections of the Air Force experts were:

- expected problems with the weaponry;
- too low of a payload, due to 3.4% of the take-off

weight being taken up by the wings' sweep mechanisms;

- insufficient fuel capacity, that is to say, deficient operational range and flight time;

- serious reservations about the configuration having the entire fuel capacity being housed directly above the engine.

(With the requested but never realized re-construction, a P1110 would have emerged)

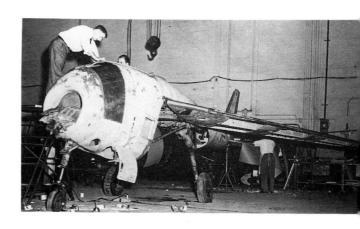

P1101 maintenance inspection in the Bell factory in Buffalo, N.Y. (1948). Work had been soon terminated. (above)

The P1101 at Bell Aircraft Works with the HeS 011 mock-up (?) and a trace of 6 x MG 151. (center)

Mid-section of the P1101 with a trace of 4 x MK 108. The result of its accident (when it fell off the transport vehicle) is viewable in the reinforced cockpit canopy. (right)

What remained was a deal with Bell for a construction of two prototypes, and the contract was concluded on July 26, 1949. These two machines, to be used as research aircraft for the feasibility of swept wing usage and for the practical use of variable geometric wing aircraft, received U.S. Air Force serial numbers 50-1838 and 50-1839.

Despite a strike at Bell Aircraft Works, a review and discussion of the wood mock-up of the P1101 successor took place at the end of the year 1949.

Wind tunnel tests at this time already showed certain liabilities in the aircraft's directional stability. To remedy this dangerous characteristic, the design received an additional vertical fin under the actual vertical tail; this was a measure that apparently was not sufficient as the crash of the second aircraft 0-1839 shows.

At the beginning of 1951, after work was under way on a few other improvements and modifications, under the influence of the NACA, the first Bell X-5 rolled out. The new prototype made its first airborne voyage strapped onto a transport aircraft, a Fairchild C-119, and the unflown bird made its way from Bell Aircraft Works near the Canadian border in a southerly direction. The end point of travel was the legendary Air Force test center at Edwards AFB in California.

On June 20, 1951, approximately seven years after the first designs for this aircraft were adopted on

the aircraft was transferred to the United States Air Force Museum in Dayton, Ohio at Wright-Patterson Air Force Base.

During the X-5 program, parts from the original P1101 were used for statistical tests, and sometime in the early 1950s, the remainder of the Messerschmitt aircraft finally went to the scrap yard.

All told, the two Bell aircraft carried out approximately 150 test flights. There was evidence that the expected high-speed and low-speed flight characteristics were successful: the German-American con-

The P1101 as a demonstration model in Bell's factory: adaptive work with the J-35 engine.

the drawing boards in Oberammergau, experienced test pilot Jean Ziegler undertook the maiden flight of the Bell X-5/0-1838. Ziegler also happens to be the test pilot who later perished in the Bell X-2. The first application with the swing wing took place a few days later during the fifth flight.

On December 10, 1951, Jean Ziegler once again carried out the maiden flight of the second aircraft, the Bell X-5/0-1839, over Edwards AFB.

After the U.S. Air Force had accepted both aircraft, an evaluation program was undertaken in which the NACA also intervened.

On October 10, 1953, this program incurred a setback due to the crash of the second aircraft, the 0-1839, as a result of inadequate pitching stability. U.S. Air Force test pilot Major Raymond Popson perished in the crash.

The remaining aircraft carried out the test and research flights until the end of 1955, and afterwards

struction illustrated a workable, though rather costly route to reach their goal of developing a high-speed swept-wing aircraft with sufficient low-speed flight characteristics.

At about the same time, the Bell X-5's counterpart from the U.S. Navy, the Grumman XF 10 F-1 "Jaguar", was carrying out some flight tests. While the swing wing mechanism worked satisfactorily for the "Jaguar's" maiden flight on May 19, 1952, the Westinghouse J 40 engine never achieved the desired output. For this reason, the evaluation program came to an end because those in charge had already had quite a nail-biting year as it was.

The experts answered the question regarding optimum angle of sweep for a combat aircraft on the basis of the available results with a value in the scale of 40 degrees on the t/4-line; Messerschmitt's aerodynamicists had already known this way back in 1943.

In the period following, Robert J. Woods, also known as the father of Bell, presented more and more proposals to the Department of Defense in Washington, D.C., for the military application of a swing-wing aircraft. But only at the end of the 1950s did the Pentagon show renewed interest and undertook the TFX Program from which the General Dynamics F-111 design emerged. The earlier Messerschmitt swing-wing studies and Bell's research aircraft represented predecessors for the F-111, and the knowledge, statistics, and data gained with the X-5 were invaluable for the design and construction of the new aircraft which completed its maiden flight on December 21, 1964.

The F-111 was just the beginning: after the F-111, throughout the West and the East a whole series of aircraft all possessing one feature emerged: wings with variable sweep. Aircraft with this type of construction today belong to the standard equipped aircraft inventory of almost every modern air force, thanks to the characteristics which were considered extremely beneficial by one Dr. Alexander Lippisch, a Professor Messerschmitt, a Dr. Richard Vogt, and an Engineer named Woldemar Voigt.

"Father and Son" —P1101 and the Bell X-5. (above left)
Bell X-5, the first prototype 0-1838. (below left)
Take-off of the first prototype during an evaluation flight. (right)

127

Immediately after lift-off: undercarriage retracts. (top)

Passing flight with 20 degree sweep setting in low-speed flight. (above center)

Passing flight with 60 degree sweep setting in high-speed flight. (left)

In horizontal flight high over Edwards Air Force Base, California. (left)

Second prototype in low-speed flight. (center left)

Landing of the Bell X-5/01839 (Second prototype, below).

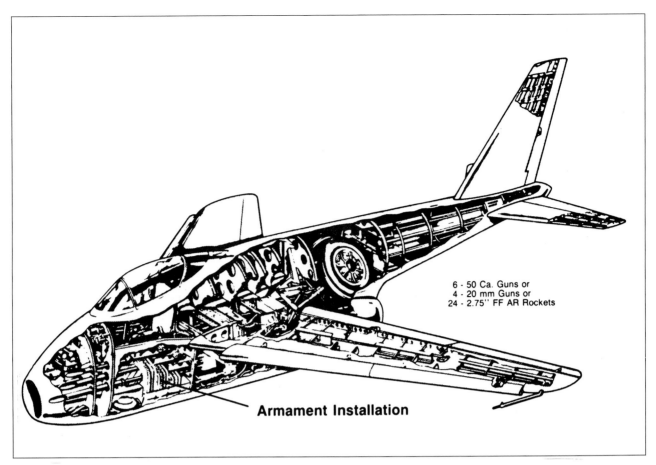

6 - 50 Ca. Guns or
4 - 20 mm Guns or
24 - 2.75'' FF AR Rockets

Armament Installation

Revell model of the Bell X-5. (above)

*Bell X-5 proposed interceptor for the U.S. Air Force:
compared with P1101 series aircraft. Source: Jay Miller/
Aerofax. (top)*

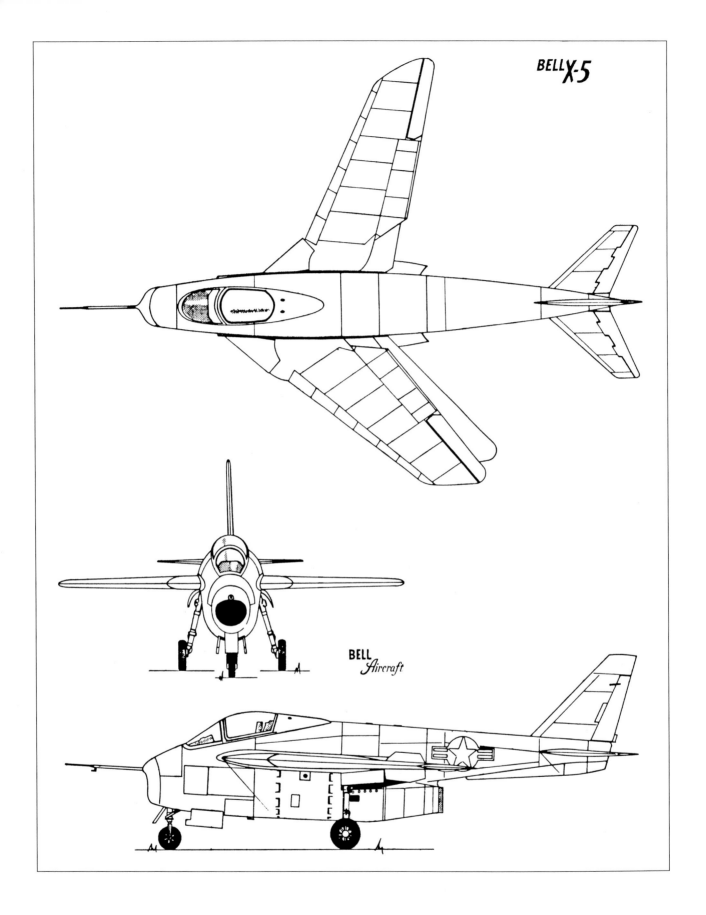

BELL X-5

BELL
Aircraft

132

Technical data of the Bell X-5
Primary function: Single-seat research aircraft for wings with variable sweep settings
Construction: Conventional all-metal made of duraluminum; climatized cockpit with ejection seat; hydraulically mechanized air brake on the fuselage nose wheel.

Length:	10,160 mm
Height:	3658 mm

Wing assembly: The sweep angle of the wing can vary between 20 and 60 degrees.

To compensate for center-of-gravity displacement when the wing is swept back into place, it (the wing) slides back a total of 686 mm. The continuous manipulation of the wing assembly takes place with an electro-motor via a synchronizing gear and a threaded spindle drive.

Wing span: Take-off setting: sweep 20°: 10210 mm
High-speed flight: sweep: 60°: 6325 mm

Wing surface:	16,26 m²

Wing assembly profile:

Root	NACA 64 A 011 with 11% thickness
Wing tip	NACA 64 a 00828 with 8% thickness
Engine:	1 Allison J-35 A-17 jet-propulsion engine with a static thrust of 2223 kp

Weight:

Empty	2880 kg
Take-off	4480 kg

Performance:

Maximum speed in horizontal flight:	1045 km/h
Maximum speed:	1135 km/h
Maximum altitude attainable:	12,800 m
Operating range:	1200 km

Three-sided view of Bell X-5. (left)

Two aircraft with a common wing root: North American F-86 and Bell X-5. (above)

The HA-300 Light Fighter

A new Beginning in Spain

The end of the war in 1945 also signified the end for Messerschmitt AG. The production facilities were demolished and the work force left behind took on the work which remained, though their first necessity was now to simply earn a living.

At this time there wasn't much left to Messerschmitt AG. The firm was under the executorship of the Americans and the former management from Messerschmitt was prohibited from entering the buildings.

Even Professor Messerschmitt himself couldn't move about freely until the spring of 1948. He was unable to dedicate himself to his chosen purpose in life, specifically, building aircraft, because he was forbidden in Germany for an unforeseeable amount of time.

It was inevitable the he eventually become involved with prefabricated housing, sewing machines, motor scooters and other similar projects. Many of his employees from the time before the collapse remained in the region and were ready for his call, or they already worked directly with him.

While the construction of aircraft in the rest of the world grew by enormous leaps and bounds thanks to German ideas and research, the authors of those forward-thinking designs themselves were forbidden to work in their chosen and inherited field in their very own homeland. Therefore, it is no wonder that several of the best aircraft manufacturers more or less voluntarily made their way to foreign lands for work.

But Messerschmitt would not have been Messerschmitt if, after the forced abstinence, a strong yearning for his very own calling in life once again didn't ring in his ear. Certainly, even at the outset, through all of his involvement abroad just as with his aircraft construction colleagues Heinkel and Dornier, the underlying goal was to resurrect their German firms.

In the beginning of 1951, an inquiry from Spain was delivered to him in regards to the on-going licensed construction of the Bf 109. Messerschmitt immediately jumped on this opportunity and traveled with his wife, the Baroness of Michel-Raulino, to Seville to Hispano-Aviacion. Though Spain had remained on a neutral course since its civil war, it was determined to liberate itself on one hand from dependence on foreign goods, and on the other to modernize itself into an industrial state connected to forward-thinking technology. With this trip, the German manufacturer made connections with the intent of undertaking aircraft development in Spain.

After his long-time co-worker and chief statistician, Professor Julius Krauss, had been employed as a consultant with Hispano since April, Messerschmitt once more visited the Spanish aircraft manufacturer in July, 1951, and afterwards summarized his impressions, experiences, and proposals in a memorandum. Therein he proposed to the Spaniards, among other things, the design of turbo-jet fighter aircraft and research in the area of supersonic aerodynamics.

This memorandum was the basis for a contract which Messerschmitt signed with Hispano-Aviacion (HA) with the consent of the Spanish Ministry of Aviation on October 26, 1951, in Munich.

Based on this agreement, on January 3, 1952, a German team commenced work in Sierra Nevada. The leader of the small group and permanent representative for Messerschmitt was Professor Krauss. Along with the team and the former Director of Project Group 1 were Hans Hornung, Mr. Ebner, and Mr. Bosch.

Following a call for bids from the Spanish Aviation Ministry, the small group, along with the Spanish technicians and with support from the Munich Bureau, developed the HA-100 training aircraft, which completed its maiden flight on December 10, 1953.

Besides the development and construction of trainer aircraft, from the beginning, Messerschmitt pursued an ambitious goal: construction of a light supersonic fighter aircraft whose performance capabilities, weight, and manufacturing costs were supposed to represent an optimum compromise.

With this design philosophy in mind, Messerschmitt, repeatedly proved that he was the master at construction of aircraft with the highest performance capabilities and the lowest possible weight. This accommodated the ideas of the Spanish Air Force's leadership.

The Spanish wanted: a small, single-seat fighter-interceptor to defend strategic points in clear weather; to attain a speed of between Mach 1.3 and 1.5; and which could be manufactured in large quantities by economically and industrially week countries.

The first information about this fighter is found in a letter from the beginning of July, 1952, from the Oerlikon Firm to Professor Messerschmitt in which the discussion concerned the installation of a "new revolver cannon" and a "compliment of rockets for a fighter to be developed by you (Messerschmitt)." The cannon under question possesses a caliber from 20 or 30 mm, and the diameter of the chosen rockets amounts to 5 or 8 cm. In this letter there is already concern expressed about a problem which would constantly accompany the future aircraft: the procurement of a suitable engine.

Further information is found in the beginning sections of a report from July 14, 1952, in which concerns about the armament and engine output of the "future jet fighter."

It was only about one year later, in a memo from July 12, 1953, when Messerschmitt instructed his co-workers, Hornung and Krauss, to build such a "turbo jet-engine fighter, through whose application of the same fuselage, either an unswept wing, a highly swept wing, a delta wing, a horizontal tail or nothing at all can be attached."

Following the two trainer aircraft with identification codes 100 and 200, the work on the jet fighter with designator P (Project) 300 was in progress.

At the beginning of 1954, Hornung and Krauss made contact with the Confederate Aircraft Works in Emmen, Switzerland, near Lucerne, with the goal of carrying out the necessary high-speed evaluations for the P 300 there.

An official order for development of the aircraft was still not available "by law", but the preparedness of the Ministry of Aviation to accept the wind tunnel costs signaled a "de facto" status which made the works possible.

With the studies from 1954, the relationship to the works in Oberammergau were still unmistakable. With these proposals the planners used a wing assembly which possessed a larger relative profile thickness on the wing tip than at the wing root.

The design of the wind tunnel mode shows the contours of the first design for the fighter project 300 (Beginning of 1954).

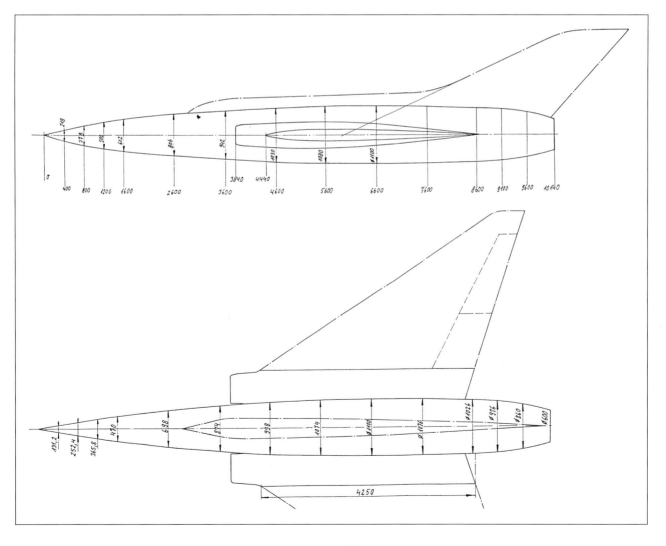

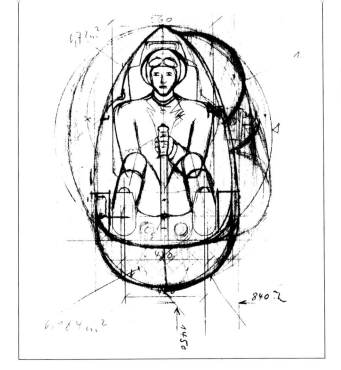

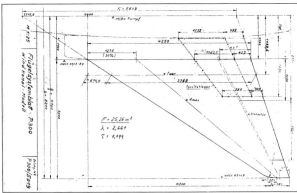

Even with the P300: An attempt for the smallest possible fuselage cross-section: Original sketch from the Spanish Messerschmitt Bureau. (left)

Wing assembly for the P300: dimensions from April 8, 1954. (below)

Full-view drawing of the P300 from October 18, 1954: of interest are the swept intakes which remind one of the Republic-105 "Thunderchief" (Maiden flight: October 22, 1955).

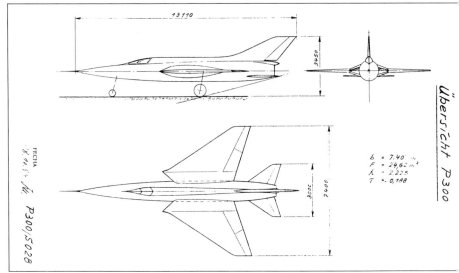

A short comparison of the principal geometric data should illustrate the differences to the P1111 from 1945:

	P1111	P300 Swept wing	P300 Delta wing
Wing span	9160 mm	8200 mm	7300 mm
Wing surface	28 m²	25.26 m²	24.52 m²
Aspect ratio	3.0	2.66	2.173
Relative Profile thickness:			
Inner	8%	5%	5%
Outer	8%	9%	9%
Spiking	0.3	0.1	0.2
Sweep	45°	50°	50°
Length	8920 mm	11,915 mm	11,915 mm
Height	3060 mm	3300 mm	3300 mm

In Seville, Hornung consistently pursued the ideas provided by Messerschmitt for the smallest possible fuselage cross-section. A sketch from the Planning Bureau clearly shows this. Just as with the later P1112 studies, a horizontal tail was anticipated for the planned fighter, though there still wasn't any clarity for the final requirement. From the P300, a proposal exists for the production series (S 028) and a scaled form for the wind tunnel evaluations where the most diverse modifications were possible: the following is known:

- Swept wing with varying wing span;

- Swept wing modified to the delta wing;
- Tail unit with high set horizontal tail;

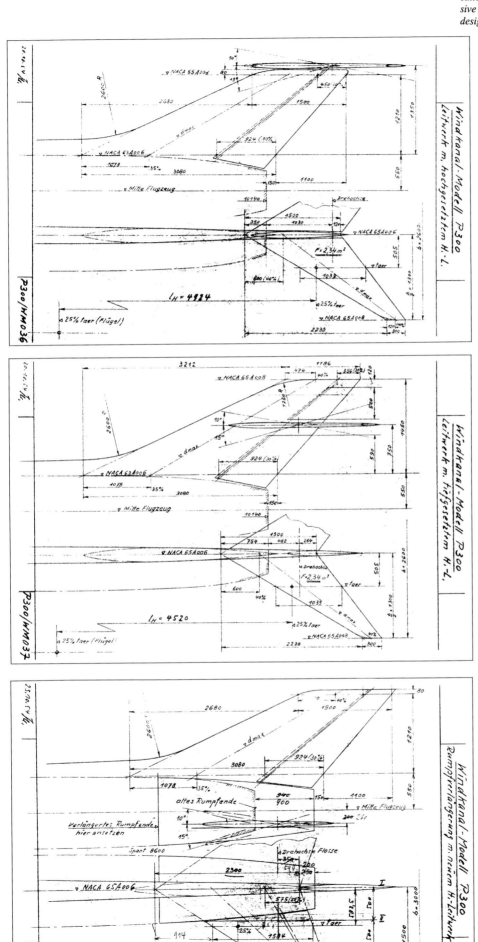

Dimensions and specifications for the planned modification on the P300 basic design: these cover the extensive wind tunnel tests for attainment of the best possible design.

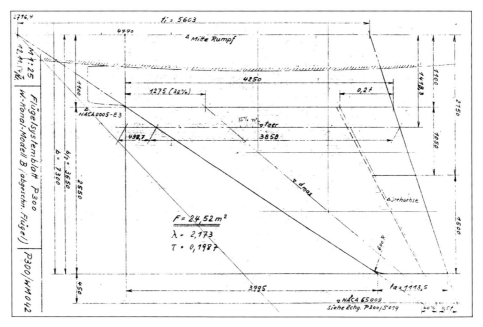

Flügelsystemblatt P300
(W-Kanal-Modell B (abgeschn. Flügel))

$t_i = 5603$

NACA 0005-E3

$F = 24,52 m^2$
$\lambda = 2,173$
$\tau = 0,1987$

NACA 65009
Siehe Zchg. P300/5079

M 1:25

P300/WM042

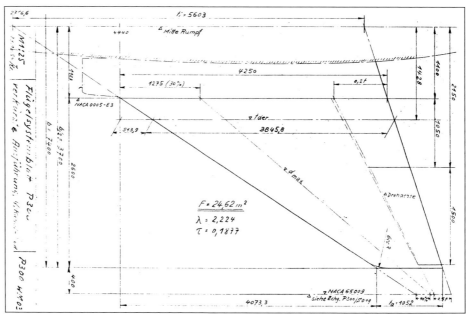

Flügelsystemblatt P300
(verkürzte Ausführung, W-Kanal -)

$t_i = 5603$

NACA 0005-E3

$F = 24,62 m^2$
$\lambda = 2,224$
$\tau = 0,1877$

NACA 65009
siehe Zchg. P300/5079

M 1:25

P300 WM02

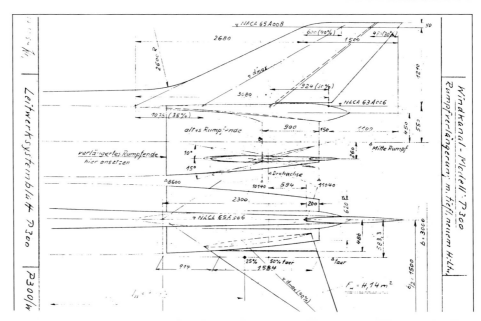

Leitwerkssystemblatt P300

Windkanal-Modell P300
Rumpfverlängerung m.tiefl.neuem H.Lkw.

NACA 65A008

NACA 63A006

altes Rumpfende

verlängertes Rumpfende
hier ansetzen

NACA 65A006

$F_H = 4,14 m^2$

P300/W

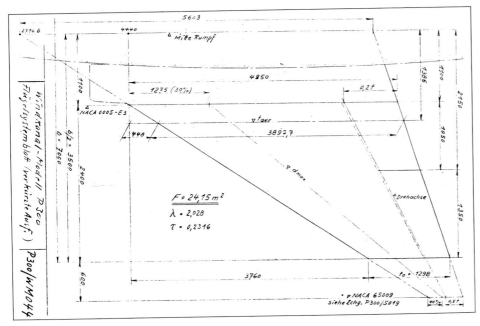

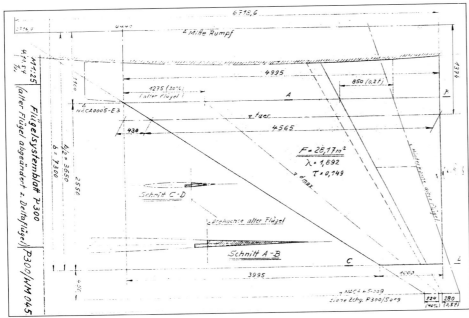

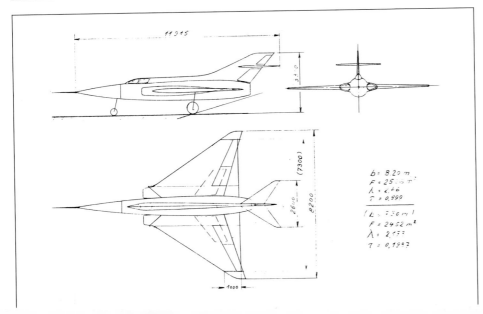

- Tail unit with deep set horizontal tail and;
- Fuselage enlargement with horizontal tail set deep on the fuselage.

After the INTA (Instituto Nacional de Technica Aeraunitica) in Torrejon, Spain, near Madrid completed the first series of tests in their own wind tunnel, Krauss and Hornung prepared for tests in Switzerland. At about the same time, Messerschmitt had a conversation with Oerlikon concerning the armament of the future defensive fighter. Again and again they discussed the questions of the engine, which, in the case of the HA-300, was never to be resolved.

At this time, Spanish aviation circles were discussing the inclusion of the American firm Lockheed and establishing, on a wider basis, the merging of the development capacities from HA and their larger competitors CASA for the independent Spanish aviation. At times, Lockheed was interested in the legal rights for the HA-300. In addition, there was a visit set to discuss the instability and other topics: Ernst Heinkel and a few of his closest co-workers halted work with the CASA and renewed old contacts.

Sometime around the middle of 1954 two-thirds of the German development teams were at work on the HA-300. Project works, statistical calculations, wind tunnel evaluations, equipment studies and last but not least, the construction preparations belonged to the most urgent tasking.

1955 brought a series of significant events, but not for the "Oficina Technica Professor Messerschmitt from Seville."

The project work on the HA-300 provided continuous wind tunnel tests with the INTA in Madrid.

The design possessed the following data at this time:

Primary function:	Single-seat fighter used for air superiority and interceptor purposes
Status:	End of 1954 and beginning of 1955
Crew:	1 pilot in a pressurized cockpit with ejection seat
Engine:	1 x SNEMCA "ATAR" 101 E with 3500 kp thrust (planned by Hornung)
Wing span:	7400 mm
Sweep:	50 degrees
Aspect ratio:	2.225
Relative wing thickness:	5/9%
Wing surface:	24.62 m²
Length:	13,110 mm
Height:	3450 mm
Take-off weight:	5000 kg
Thrust load:	1.43 kg/kp
Maximum wing load:	203 kg/m²
Maximum speed:	1400 km/h
Armament:	2 x 20 mm Oerlikon cannons and 2 x Oerlikon rocket cannons with at least 24 7.5cm rockets

Following evaluation of the wind tunnel tests, which led to a modification of the wing assembly profile and a reconstruction of the intakes, construction work on the forward fuselage section could begin in Seville; and simultaneously, construction of the mock-up could also begin.

Messerschmitt confidently hoped to bring the fighter to the skies during the year 1957. Though even to him it was quite clear that this opportunity was essentially a difficult undertaking in comparison to the HA-100 or HA-200. For example, he expressed the following on March 21: "The aerodynamics of the HA-300 are a very tricky endeavor and we must do everything to avoid any sort of setback in regards to construction and testing." And on April 7, 1955, he wrote: "The HA-300 will still cause us many headaches." Therefore, it is no wonder that Spanish officials harbored doubts as well that the HA-300 could be built in Spain without difficulties.

While the concerns of the Spanish revolved more around finances, the wind tunnel tests on-going in Emmen got closer to the doubts of the correctness of the entire concept.

In a German-Spanish discussion on May 24, concerning the further direction of development of the HA-300, the following would be established in modification of the concept up to that point:

- The fuselage design must be usable for both a delta wing and for an unswept wing; where the swept wing (delta) has a leading edge sweep of 45 degrees and a thickness of 5%, the unswept tapered wing would be equipped with a tapered shape from approximately 2:1 and a thickness from 4-5%;
- the aircraft must also be able to be built as a two-seat design;
- and the power plant should now employ the British Armstrong Siddeley "Sapphire" 6.

Obviously much was still up in the air. And in May of 1955, the Federal Republic of Germany finally got back its complete air sovereignty from the allied powers as well. Aircraft construction took on another dimension as entrepreneurial activities, and for Professor Messerschmitt this meant that, with certainty, a further limitation on his involvement on the Iberian peninsula.

Ernst Heinkel, who wouldn't let his activities in Spain slow down following his visit in the summer of 1955, had an understanding with the CASA and he worked in Stuttgart on a light-fighter design which apparently mirrored the ideas of the leaders of the Spanish Air Force more so than those of the somewhat heavier Messerschmitt aircraft.

P300 model according to the status from the end of 1954/ beginning of 1955 with the enlarged fuselage and low-set horizontal tail unit (WM 043) in the wind tunnel of the Confederate Aircraft Works in Emmen, Switzerland.

The delta-winged fighter HE C-101 from the Heinkel Chief Engineer Siegfried Günther was based on the experiences he had gained with the design of a fighter-bomber for Egypt.

Although the cooperation between Heinkel and the Spanish locations had broken off, Messerschmitt reacted immediately. On July 13, he declared: "that I'd be prepared as well if the Ministry of Aviation wanted to prove the point of whether or not the aircraft design should indeed be smaller than we see it at this moment."

The big aviation event in Seville was in the summer of 1955 as the maiden flight of the Spanish jet, the HA-200.

On 12 August, before the piston engine driven HA-100, developed as a two-seat trainer aircraft by Messerschmitt and his co-workers, climbed into the blue skies over the Sierra Nevadas, there was still another bitter disappointment for the Spanish and for Messerschmitt: The Federal Republic of Germany had decided a month prior to select the French Fouga "Magister" for the standard trainer of its new Luftwaffe, and not the HA (Me)-200.

For the HA-300, whose urgency was once more uttered by the Minister of Aviation, the plans went in the direction favorable to Messerschmitt. The "Study Bureau", strengthened in size with a few more employees, worked in Seville and in Munich on the "small" design.

There also arose a considerable number of proposals, wind tunnel models, and other unfulfilled ideas.

An aircraft without a horizontal stabilizer and with delta wings and trailing edge flaps came about.

Messerschmitt proposed testing of a glider without an engine.

The new design philosophy would be visible in the specifications from November 23, 1955:

The HA-300 proposal for a design with the request for reply by December 1, 1955.

The switch-over of the HA-300 project, in so far as the net weight is concerned, to the favorable Bristol Orpheus engine, and thereby to a real light fighter, would appear desirable to decide on the application of this type of aircraft.

The geographical situation in both Germany and Spain, i.e. their coastlines, allowed considerable time from the moment for the reporting of an attack until defenses would be warranted. For fighters to patrol the airspace, however, a large expenditure in procurement is required and they are difficult to keep in operation. Therefore, the HA-300 was to be the definite interceptor and the loophole between aerial surveillance and surface-to-air missile defenses. The following ideas particularly should be investigated:

Shortest climbing time to 15,000 meters
Good fire power even with little ammunition with the 2 Hispano Suiza 30 mm.
Standard equipped gyroscopic gun sight; servo control system without autopilot; radio communications connection and range finder.
Fuel in the interior for suppressing, climbing to 15,000 meters at 100% power, hour dog-fight or 5 minutes with afterburner, 150 km landing approach to the next airfield.

With this configuration, the following weights are commensurate:

Empty weight	2206 kg
Disposable load: 300 rounds	180 kg
Fuel	950 kg
Pilot	85 kg
Total weight of:	3421 kg

For a take-off weight of 3300 kg after suppressing, there is, for middle climbing speeds, between 0 and 5000 meters about 100 meters per second; between 5000 and 10,000 meters about 70 meters per second; between 10,000 and 15,000 meters about 32 meters per second, and thereby a total climbing time from 0-15 km of 290 second or about 5 minutes.

The fuel capacities above are sufficient for a flight route totaling 500 km. If the HA-300 is called upon for a dog-fight or to attack ground targets, an additional payload of 1000 kg to 1100 kg fuel, in releasable exterior tanks, must be provided for the return flight. In this instance, the aircraft could still be a valid air surveillance fighter. This payload configuration should not, however, be the design used in dog-fights and supersonic flight for calculated landings.

Seville, November 23, 1955
M. Schäffer
Distribution:
Professor Messerschmitt
Professor Thalau
Herr Hornung
Herr Madelung
Herr Schäffer

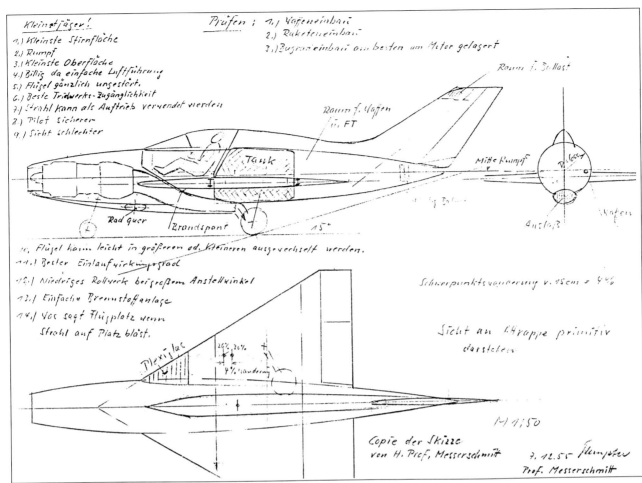

Light fighter (small fighter) proposal from Professor Messerschmitt. (Copy of a sketch from December 7, 1944).

As the hand-written note brings to light, Messerschmitt was in agreement with the proposal: however, typically for him, he considered the structural weight of 2206 kg to be too high.

Sometime later a work plan would be drawn up in which it is mentioned that the proposed glider aircraft from the AISA Firm was to be built. The deadlines are allocated on short notice, and the entire conception must be finished in January 1956 in order to be able to provide the modified project to the Ministry of Aviation.

Simultaneously, the leaders were worried about getting new material for the necessary supersonic tests at the Confederate Aircraft Works in Emmen. Problems already stood out with the financing of the project.

In Spain's Ministry of Aviation, Egypt's increased interest allowed the hopes of finding someone to cover the development costs to grow.

The Spanish government, or its Ministry of Aviation, publicized the first-ever cooperation and the commissioning of Messerschmitt for design of the turbojet fighter development, HA-300, in the daily newspapers.

The emphasis in development work once again shifted, increasingly, from Munich to Seville in increasing measures; extensive supplementary wind tunnel tests were planned and carried out and a memorandum drawn up in which the HA-300 was finally defined as a light fighter and light combat aircraft weighing from 3-4 tons. Looking at aircraft with similar power plants and similar applications (the Fiat G-91 or Folland "Gnat", for example), the HA-300 distinguishes itself, most importantly, through its (theoretically) higher flight speeds.

The original text of the memorandum from February 8, 1956, gives substantial insight into the development stage reached at that time, the performance capabilities, the construction type, and further action undertaken:

Memorandum concerning the P300 supersonic fighter design

Introduction:
The first work on the P300 project (fighter with

supersonic speed in horizontal flight) began near the middle of 1953.

For the engine, the most modern engine of its day, like the Avon and Sapphire, was planned for this aircraft conception. With the knowledge available at this time in regards to high-speed aerodynamics, it was believed that this was the only engine powerful enough to attain supersonic speeds in horizontal flight. Too much was still unknown of the high-thrust light engines, such as the Bristol "Orpheus", to align the entire project with these engines. Design models with powerful engines were tested in subsonic and transsonic wind tunnels. The results of the measurements confirmed that supersonic speeds are possible. And confirmed further, that with the new light engines and the knowledge gained of the "Area Rule Concept" by slightly affine reduction (based on the lower engine weight, smaller dimensions and also lower total fuel consumption), the same performance can be reached.

Through application of the "Area Rule Concept", it is now possible to decrease the high resistance in the Mach 1 area to a tolerable level and therefore get by without an overpowered engine which was necessary in overcoming this drag peak.

In the fall of 1955, this knowledge and experience, and especially the economical considerations, led to the transfer from this design (weight class: 5500-6000 kg) with massive engines, to the weight class of 3000-3500 kg, with the light engine Bristol "Orpheus", i.e. switched to a light fighter.

Tasking

This aircraft is designed as a single jet engine light fighter (fighter with the lowest construction expenditure) with supersonic flight speeds in horizontal flight in high altitudes, and is used first and foremost as an air defender with a flying radius of approximately 1000 km at a sustained altitude of 11 km with maximum throttle.

The tactical missions of the patrolling, aerial surveillance fighters (flight time approximately 2 hours at altitude of 11 km) and escort aircraft (flying distance approximately 1500 km at altitude of 11 km) are ensured by the use of exterior, jettisonable fuel tanks.

The aircraft may be used in ground strikes with external loads of up to approximately 600 kg, alternating with supplemental weapons or expendable loads with a flying distance of 400 km in altitude of 1,000 meters.

Two development stages are foreseen:

Stage 1 a: With today's production series engines without afterburner at 11 km altitude, to reach a horizontal flight Mach number of at least 1.25. Service ceiling no lower than 15,000 meters.

Take-off and landing distances (of no more than 1200 meters) for maximum operational weight. Supplemental afterburner to increase flight characteristics is to be considered.

Stage 1 b: Supplementary equipment, especially effective lift and climbing aids, such as the wing blower system, while keeping expenditure down, for substantial improvements of the take-off and landing characteristics, as well as high-speed flight characteristics.

A substantial increase in operational capabilities such as aerial surveillance, escort fighter, and ground attack aircraft, will be attained.

Details of the tasking

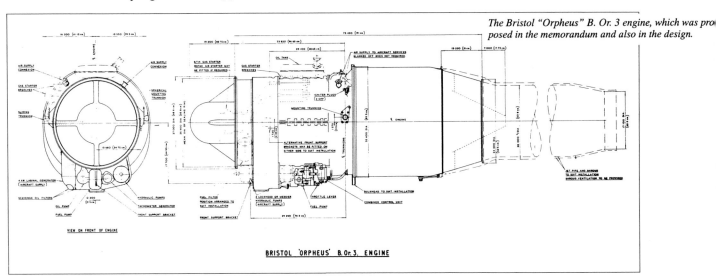

The Bristol "Orpheus" B. Or. 3 engine, which was proposed in the memorandum and also in the design.

BRISTOL 'ORPHEUS' B. Or. 3. ENGINE

Engine:

The Bristol "Orpheus" is the planned power plant. This engine has a low structural weight and performs favorably with relatively high thrusts and fuel consumption. According to data from the manufacturer, the present engine with 2200 kp static thrust is still in the development stage, i.e. it can be increased, and in the next stage performs at approximately 2600 kp static thrust. With afterburner, an additional static thrust increase of up to 40% is attainable.

In the follow-on development stage the engine should be able to increase its lift capabilities.

Armament:

The application as a "light fighter" in place of interceptor/aerial surveillance and escort aircraft would result in a reduction in the armament weight.

An alternative weapons choice is considered desirable for effective engagement against aircraft, i.e. either 2 x Hispano Suiza 30 mm with 120 rounds each or 20 x 5 rockets (fired from cannons) built into the aircraft fuselage.

For ground operations in the first stage of development with decreased flight capabilities, the following is intended: an additional 4 x Hispano Suiza 23 mm firearms or supplementary rockets, or releasable load up to a total weight of approximately 600 kg.

Standard equipment:

Reduction to the extreme, most necessary equipment for all weather flight.

Flight performance:

Maximum horizontal flight speed at an altitude of 11 km without afterburner and with a Mach number no lower than 1.25.

Climbing time to 15,000 meters not to exceed 5 minutes with 2600 kp static thrust.

Take-off and landing:

In the first development stage for all operational uses, the take-off and landing distances are not to exceed 1200 meters.

Construction for implementation of this configuration with the P300 design:

A general description of Development Stage I after the phase from January 1956 (not yet finalized and modifications are at hand):

Thinner delta wing with aspect ratio of approximately 1.9 with a wing surface of approximately 17 m2 and a wing span of approximately 5.6 meters. Total aircraft length: approximately 10.0 meters.

Fuselage breakdown from front to rear:
Fuselage nose consists of a nose wheel and complete electric mechanism.

Cockpit: pressurized and with ejection seat.
Forward fuel tank: (approximately 680 liters).
Weapons installation: Below and behind the forward fuel tank.
Engine: With airflow from both sides of the fuselage.
Main undercarriage wheel: In retracted position under engine.
Rear fuel tank: (approximately 680 liters).
Afterburner and exhaust stack.

In observance of the smallest fuselage cross-section surface, slimness and fulfillment of the "Area Rule Concept" is to be adhered to.

Weights: (preliminary data) with "Orpheus" 3. Empty weight (with 2 x 30 mm weapons) 1950 kg
Additional loads:

Ammunition		150 kg
Fuel and lubricants		1100 kg
Pilot	85 kg	1335 kg
Total weight		3285 kg

Performance:
Note:

To deliver exact and guaranteed flight performance characteristics at this point is impossible, since insufficient data are available in the trans-sonic and supersonic flight regions in Mach and wall effect in relationship to resistance.

The performance assessments were carried out dependent on the available trans-sonic measurements and theoretical calculations in the supersonic speed range.

Maximum horizontal flight speed:

With 2200 kp static thrust
at altitude of 6 km: speed = 1140 km/hour or Mach ~ 1.00
at altitude of 11 km: speed = 1360 km/hour or Mach ~ 1.28
at altitude of 15 km: speed = 1070 km/hour or Mach ~ 1.00

With 2600 kp static thrust
at altitude of 6 km: speed = 1490 km/hour or Mach ~ 1.31
at altitude of 11 km: speed = 1490 km/hour or Mach ~ 1.40
at altitude of 15 km: speed = 1120 km/hour or Mach ~ 1.05

With 2600 kp static thrust and 40% afterburner
at altitude of 6 km: speed = 1870 km/hour or Mach ~ 1.65
at altitude of 11 km: speed = 1920 km/hour or Mach ~ 1.80
at altitude of 15 km: speed = 1760 km/hour or Mach ~ 1.65

Cruising speed at altitude of 11 km with 95% thrust (maximum throttle) with 2200 kp engine static thrust
Cruising speed = 1330 km/hour or Mach ~ 1.25

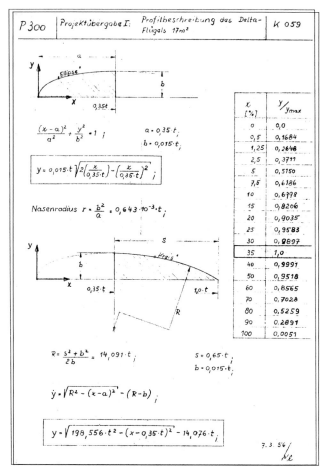

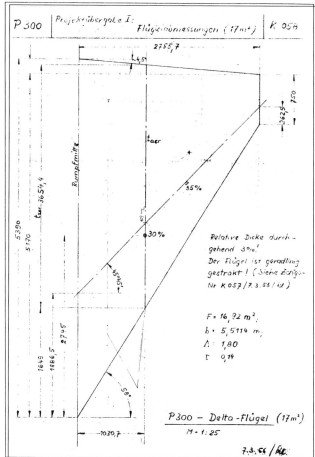

The HA-300 wings developed: Project Handover I from March 7, 1956.

with 2600 kp engine static thrust
Cruising speed = 1460 km/hour or M ~ 1.37

Climbing speed with full throttle:
(At Mach ~ 0.9)

with 2200 engine static thrust
at sea level w = 110 m/second
at altitude of 11 km w = 33 m/second
at altitude of 15 km w = 8 m/second
with 2600 engine static thrust
at sea level w = 151 m/second
at altitude of 11 km w = 50 m/second
at altitude of 15 km w = 14 m/second
with 2600 static thrust plus 40% afterburner
at sea level w = 261 m/second
at altitude of 11 km w = 80 m/second
at altitude of 15 km w = 31 m/second
Climbing time with full throttle:
with 2200 kp static thrust
at 11 km t = 2.85 minutes
at 15 km t = 6.50 minutes
with 2600 kp static thrust
at altitude of 11 km t = 2.00 minutes
at altitude of 15 km t = 4.50 minutes
with 2600 kp static thrust plus 40% afterburner
at altitude of 11 km t = 1.10 minutes
at altitude of 15 km t = 2.30 minutes

Take-off and landing:
Take-off distance with 2200 kp static thrust
Distance = 460 meters
Landing distance with 2/3 fuel and with ammunition
Distance = 1080 meters

Flight distances and flight time:
At maximum throttle (2600 kp static thrust)

Fuel 1100 kg

Altitude	Flight Distance (km)	Horizontal flight time/hours
sea level	400	0.33 hours
6 km	630	0.45 hours
11 km	1070	0.76 hours
15 km	1560	1.45 hours

Equipment:
Bristol "Orpheus" B-3 engine with afterburner
Baker M 4 ejection seat
VHF radio
Range finder (supplementary)
Gyroscopic compass
IFF system
Gyroscopic gun sight
Flight and engine aerial surveillance equipment
Fuel control
Air pressure and climate control system for cockpit
Armament: 2 x Hispano Suiza 30 mm
Ammunition: 240 rounds
In place of weapons:
2-3 cannons with 20 rockets

Munich, February 8, 1956
Professor Dr. Engineer Willy Messerschmitt

In early 1956 the engineer leaders were in agreement over the final configuration of the wing assembly. As is apparent from Project Delivery I from March 7, 1956, the delta wing was now designed with a continuous relative profile thickness of 3%. Such a wing assembly in the 1950s represented a special type of risk: the Lockheed F-104 "Starfighter" possesses a thickness-to-chord ratio of 3.4%, while another high-speed jet aircraft of this era possessed a wing thickness of 4 and 6%.

This merely shows that, despite the sometimes adverse circumstances of years past, Messerschmitt had not only made the connection, but he also pursued his interests with the intent of continuing on much further.

By the middle of the year there were up to 18 German experts working on the new HA-300 project. The jobs were characterized by extensive weight comparisons with different engines, allocations and cost breakouts, and diverse wind tunnel evaluations in Emmen and in the Upper Bavarian Grainau. In the meantime, by the end of the year, operations were once again taken up in the overhauled wind tunnel of the INTA.

Messerschmitt, at this time residing predominantly in Munich, founded the Flugzeug Union Süd (FUS) with his old rival Ernst Heinkel in August, 1956. The purpose of this company, ironically enough, was the licensed construction of the French Fouga "Magister" for the Federal Republic of Germany's Luftwaffe.

While German engineers were already working on realizing the light supersonic fighter in Spain, on November 2, 1956, the Ministry of Defense in Bonn called for an "interceptor" for the Federal Republic's Luftwaffe, since the German aircraft industry had proposals under construction at that time.

Messerschmitt contributed with a HA-300 proposal with forward-set horizontal stabilizer.

At the same time in Seville, and above all from Professor Ruden and the Spanish engineer Rubio, this aircraft shape designated as "duck construction type" was strongly favored. Professor Ruden proposed the testing of the present configuration as a development stage for glider presently under construction and, with positive results, also for the HA-300.

For his proposal, similar to the later SAAB J-37 "Viggen", Messerschmitt could hardly find any supporters. This time Ernst Heinkel had the leading edge with his He 031 "Florett" design. However, the powers in Bonn gave precedence for procurement of aircraft from the allied foreign countries, and so followed through with the initial delivery to the Luftwaffe with the Republic F-84 F "Thunderstreak" and the Lockheed F-104 G "Starfighter."

Besides the continuous modification proposals, the project work on the HA-300 also endured a revitalization which occurred between the German and Spanish engineers.

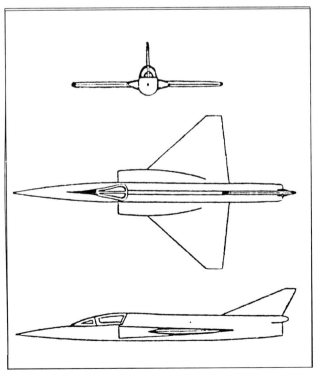

A sketch which appeared in the German press in 1957 of the Me-300, as it was then, HA-300 and/or XC-6 designated light fighter project.

Such was written in a letter from Seville: "The situation with the 300 becomes more critical every day, since Rubio made a counter-proposal with the fuselage's underside intakes with the same construction (the nose undercarriage hadn't been investigated at this point) and for the floor plan, he was better off with this design. Rubio was very clever to present the design to influential people as a fine wood model and the question then became: why can't this design be built. Urgent. Prelude to a battle over prestige!"

Apart from that fact, construction of the glider or sail aircraft was practically complete by the end of 1956. And its wing assembly, to be constructed from wood but not with 3% thickness, was almost under construction.

In comparison to the data from the beginning of the year for the HA-300, now the Bristol "Orpheus" Br. Or. 11 engine with 2610 kp thrust was intended. The empty weight rose from the former 1950 kg to 2335 kg and the total weight from 3285 kg to 3485 kg.

In increasing degrees the international press became interested for the German-Spanish project. Even in German technical journals, the first design drawings and sketches surfaced with the aircraft designated as either HA-300, Me-300 or XC-6.

The perceptible money and currency deficiencies of the Spanish government had a crippling effect on the work in Seville. This was evident not only by the difficulties with wages for the "Spanish" Messerschmitt employees, but the circumstances also halted the manufacturing preparations to a large extent for the two prototypes of the HA-300 that had been ordered in the meantime, and for which the first parts were already available for the test phase.

For the same reasons, the wind tunnel evaluations were stored in Grainau and Emmen and they sent back equipment, measurement systems, processing equipment and other similar equipment for the HA-300 test construction, or without further ado, canceled the orders.

In June, 1957, in order to be able to carry out short-term aerodynamic tests they considered construction of a free-flying model, shot with the use of a catapult, in a scale of 1 : 10. The idea was still there.

Despite the strained financial situation the development work appeared to be energetic. Messerschmitt and his leading employees, among them a nephew and later MBB (Messerschmitt-Bölkow-Blohm GmbH, Flugzeug- und Raketenkonzern) Group Chief Gero Madelung, concerned themselves with the details of equipment, the materials, and the production and manufacturing of the supersonic light fighter.

Somewhat more obvious were the efforts for usable testing and measurement results to win over the estimate documents for the stress distribution of the wings, tail unit, ailerons, (combination of horizontal stabilizer and aileron) and the flaps. By the same token the planning called for the most powerful thrust from the "Orpheus" variant, of the Bristol Or. 12 with a static thrust of 3090 kp (3630 kp with afterburner) and thereby the corresponding increase in speed to over Mach 1.6 not only by widening the intakes by enlarging the maximum fuselage cross-section but also by new, cost-intensive wind tunnel tests.

Also hard at work on the HA-300, the Munich development team was, by mid-year, released by Messerschmitt AG and worked as an independent "study group" on a study contract for the Ministry of Defense in Bonn. A portion of the German employees of the "Oficina Technica Professor Messerschmitt Sevilla", who had a three-month hiatus in Germany the summer and spent their vacations during this time, were restricted from this study contract and therefore, suspended from working on this contract with the Oficina Technica.

This was just the beginning of the decreasing involvement of personnel from Professor Messerschmitt and his team in Spain.

Until the end of the year this was reduced: from twenty employees to only ten remaining for the continuing work in Seville.

Nevertheless, the drawing up of factory designs proceeded relatively well at the beginning of 1958. In February, the cockpit got its final design, and at the beginning of March the work appeared to be proceeding to everyone's general satisfaction. A letter from Messerschmitt to Madelung gave some information of the constructive developments and progress with the wings, with the flaps, the fuel system, the hydraulics, and climate control system, and so on: it stated further: "The cockpit canopy became ridiculously simple without hydraulics and electronics. Fastening of the fuselage rear (over the steel frame) is so magnificently resolved that one can be happy about it. The electrical system has become simpler, as we've returned to the old system which the F-100 possessed. The undercarriage is also now clear, and the rotation has become simpler and retraction lies clear and close to the underside of the frontal wall. What we're lacking are aerodynamic tests, etc. etc."

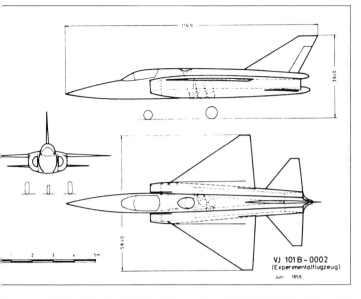

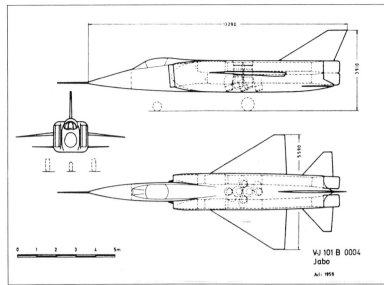

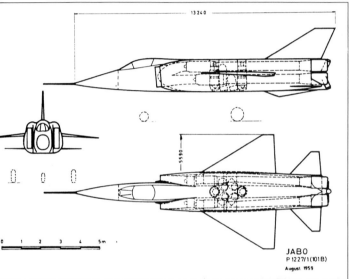

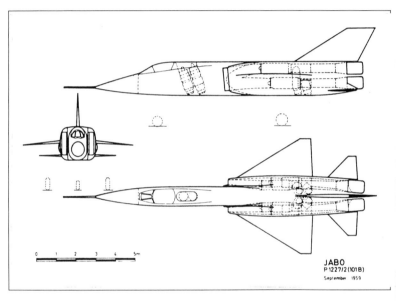

The influence of the development work in Spain on the activities in Munich: vertical take-off and fighter-bomber aircraft proposals in the summer of 1959; and also herein a quickly realized "Experimentation Aircraft" based on the HA-300. (from top-left to bottom-right)

The sparser and sparser correspondence between Munich and Seville, along with the reduced Messerschmitt team in Seville, illustrates the lowering interest of the Spanish on the development work by Willy Messerschmitt. The reasons probably lie in the already noted strained financial situation of the Spanish government and in the increasing burden on aircraft production factories with orders from the Americans. These orders at HA firstly affected the Lockheed T-33's under repair or being assembled and further affected the licensed construction of this aircraft type for the Spanish Air Force and clearly showed the American influence.

On February 23, 1959, Heinkel, Bölkow, and Messerschmitt founded the Development Group-South (Entwicklungsring Süd—EWR-Süd) in Munich. Within this framework the planning and design of vertical take-off combat aircraft for the German Federal Republic's Luftwaffe ensued. Work on the HA-300 strongly influenced the designs of Messerschmitt's Study Bureau. Above all, the intended "Experimentation Aircraft" VJ-101 B-0002 is a direct derivative of the Spanish light fighter. The advantages of proceeding in this method are obvious: the HA-300's wing assembly had already been tested in the wind tunnel and this available data would save time and money. Furthermore the completed construction formed a safe basis for design.

In the meantime, in Spain one thing was once again unclear: the HA-300 development was decidedly influenced by the looming impossibility of a timely discussion of the Bristol Br. Or.12 engine.

In addition, the wind tunnel tests brought doubts about the stability of the aircraft, and moreover they observed a dangerous pitch of the control element and stabilizer fin for flutter. After the designs were completed as far as fuselage and equipment were concerned and the largest portion of the equipment ordered, then construction of both prototypes was discontinued.

Meanwhile, the designed glider, a mixture of many construction designs, finally stood ready to begin evaluation of the low-speed flight characteristics of the supersonic fighter. On June 25, 1959, in San Pablo, following the taxiing test behind a jeep, the glider, in tow of a CASA C 2.111 (Spanish configuration of the He 111) took-off. Immediately after lifting off the ground the sail plane showed detrimental instabilities which forced the pilot to quickly abort. And thereby the first and only attempt to fly the aircraft designated as HA-300 P was ended.

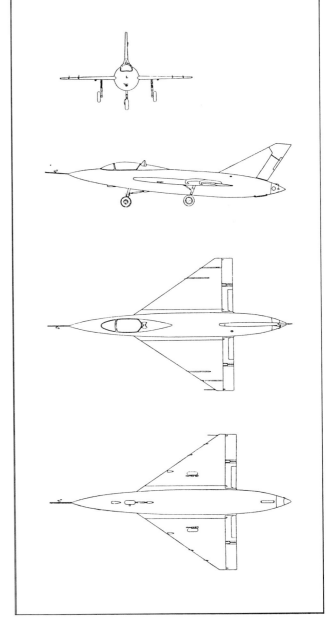

Technical data for the glider:

Wing span:	6150 mm
Wing surface:	20.0 m² l = 1.89
Sweep of leading edge flap:	57.7 degrees
Length:	10,200 mm
Gross weight:	1250 kg with water ballast and two balancing tanks

Messerschmitt's contract for consultant work also ran out during this same year. The remaining employ-

Glider to evaluate the flight characteristics of the design.

ees of the "Oficina Technica Professor Messerschmitt" were returned from Seville, with the exception of engineer M. Blümm. Since Messerschmitt had acquired 27% of the shares of Hispano Aviacion, Blümm remained on the Iberian island as his middle-man.

At the end of 1959, Spain sold the completely defined HA-300 project, along with all documents, to the United Arab Emirates who then took on the second attempt to create a combat aircraft from the design.

Finale in the Nile Valley

At the beginning of the 1950s, Egypt, in collaboration with and under management of Ernst Heinkel and his team, began to build a fighter-bomber. Approximately 30 kilometers south of Cairo, in the Nile Valley on a piece of land left behind by Britain, there was an airfield and manufacturing plant, and it was there that an aircraft and engine production plant emerged. In Stuttgart, Heinkel had outlined a pure delta-winged aircraft and constructed an engine for the aircraft designated as He 011. This engine had a static thrust of 5880 kp and 8100 kp with afterburner. In 1956, after twenty years of work, the entire operation fell through, and for Egypt, all that remained was an ornamental airfield with three new buildings and a half-finished engine test bench as a reminder of a considerable sum of money literally stuck in the sand.

The acquisition of the HA-200 license and the HA-300 documents during 1959 shows that, with this failure, the country's air force leadership had in no way buried their dreams of an independent, national aviation industry along the "Holy Nile." Co-decisive in this purchase was certainly the news that Israel intended to procure the French Mirage III. Just as other lands from east to west had done following the war, the Egyptians turned to German specialists for aid in building up their aviation industry, as the Germans, until long after the war, did not have the opportunity to prove their abilities and talents in their very own homeland.

On the initiative of the Egyptian-Swiss businessman and engineer, Hassan Kamil, Messerschmitt closed a deal with those responsible in Cairo. The contract, at its core, looked like this:

- development of a small fighter-interceptor with approximately 4 tons of take-off mass, designed on the basis of the HA-300 for Mach 2.2 speed;
- familiarization and training from native technicians as a basis for a future industry;
- and the completion of the initial production of the supersonic fighter.

Almost simultaneously the founder and president of the Swiss construction company (MECO), Hassan

Kamil, recruited a total of 470 technicians from Spain, Austria, Switzerland, and Germany to support construction of the aircraft. Even though the ambitious undertaking had motivation and the working conditions were thoroughly in order, when viewed superficially, even the requirements for the plan were not especially favorable:

Egypt was a North African state which had just made steps to make its way out of the feudal age into the modern world and for them, high technology was lacking from the entire infrastructure. This was most painfully evident in the field of electronic machinery for manufacturing, testing, and even for aircraft. On one side there was a long-standing conflict with Israel, who could engage in a war once again at any moment; moreover, the problems with the western world had piled up, and with all the affluence, relations with the neighboring Arab feudal states weren't exactly the best.

In addition, the leadership came to Cairo with time restraints since the Israelis planned to incorporate the Mirage III in 1962, and in comparison to the Egyptian MiG-19, the Israelis had a dangerous military-technological advantage.

The planners met the first blow at the beginning of 1960, as the English engine manufacturer Bristol Aero Engines declared the discontinuation of the work on the "Orpheus" Br. Or. 12. The British arranged for the continuation of the working conditions which couldn't be accepted by either interested party (India or Egypt).

And so the question of the engine was once again totally up in the air.

In April 1960, Professor Messerschmitt invited the well-known engine builder, Ferdinand Brandner to Munich for a talk. The Swiss Brandner had worked very successfully earlier for Junkers and for the Soviets. Brandner was suitable for the task and so an agreement was reached in the summer of 1960 with him to bring about an engine in mass production for the HA-300 by 1965 with the dimensions of the Br. Or. 12. Brandner promptly got to work on resolving the personnel and organizational problems.

In the relatively short time span from 1960 to 1964, in a multi-national show of strength, an aircraft and engine factory emerged from the Nile Valley, which encompassed an area of approximately 120,000 m².

An airfield sized approximately 4.5 km² belonged to the complex. The most striking designs of this extraordinary achievement were the four large engine

engine test benches, next to the spacious airfield and which visibly towered over the manufacturing plants, repair stands, and test systems.

On July 25, 1962, President Nasser officially opened the first manufacturing plants. One of the first programs in Heluan was the licensed construction of the HA-200, which the Egyptians designated as "Al-Kahira." The engine of the "Al-Kahira" came from Heluan: Ferdinand Brandner and his team developed the engine in close connection with the French Turbomec "Marbore", and as early as June, 1962, the first of a total of 170 rolled off the test stand.

The aircraft construction, under the technical leadership of Messerschmitt, was being directed in the hangars of Factory Number 36 during the 1960s, while the technical and organizational director of engine construction, Ferdinand Brandner, along with his co-workers, moved into the completely new Factory Number 135, which was continually under construction itself.

The HA-300 was now being developed and constructed in Munich by the "Messerschmitt Study Bureau" and in Heluan with the participation of a few Germans.

From the flight tests with the glider in the early summer of 1960, and from the final wind tunnel evaluations and the somewhat modified tactical requirements, which came with the move from Spain to Egypt, Messerschmitt drew his conclusions. Together with engineer Blümm he modified the delta wing somewhat to suit him and designed a horizontal stabilizer for the HA-300. The horizontal stabilizer created somewhat more resistance which had a relative effect on the maximum speed, and simultaneously permits the larger distance from the aircraft center of gravity and the associated lower control surface deflection to create a better performance in turning, a higher rate of climb, and last but not least, it means a force-moderate release of the wing. The horizontal stabilizer of the HA-300 is undampened in order to achieve sufficient control surface effect in supersonic flight. Since Messerschmitt, likewise, provided this component for resistance and weight reasons with a very light profile, he evoked a danger which he had to correct later in connection with the built-in control: the tilt of the flutter of the horizontal stabilizer.

In September 1960, the Swiss Federal Aircraft Plant in Emmen carried out the subsonic measurements of the modified HA-300 model on a scale of 1:3. The British took over the supersonic measurements in the summer of 1961 in the wind tunnel of the Bedford Research Institute. Both series of evaluations confirmed the correctness of the previously carried out developmental steps. The director of the laboratory at that time, Mr. Hill, even spoke of the best results since the existence of the institute. The later flight test nevertheless showed that theory in the wind tunnel and actuality in the air can be two different things when it comes down to details. Messerschmitt himself tried to order an "Experimentation Aircraft" as quickly as possible

At this time in the Federal Republic, the VJ-101 Program was already ongoing. The final goal of this program was the design and construction of a supersonic vertical take-off and landing fighter aircraft. Messerschmitt, who was only occasionally in Heluan, had his firm heavily involved in this project, along with Heinkel and Bölkow in the context of Development Group-South (EWR).

Messerschmitt wanted, if at all possible, to fly the HA-300 in 1963 in Egypt. This goal was to be reached with the E-300 engine, whose first bench test run was planned for the middle of 1963. But the goal was not to be attained. Against the avowed will of Brandner, 25 British "Orpheus" 703 S-10 engines with a static thrust of 2200 kp were imported. It was obvious that the HA-300 would never, ever demonstrate its full potential with this engine. The British engine had some slight similarities with the E-300, right up to the nearly identical dimensions:

- the Brandner engine, with its afterburner, had almost twice the length of the "Orpheus";
- the dry weight amounted to 2.3 times more, and;
- the substantially higher air volume of the E-300 required, with installation of this engine, another configuration for the intakes.

In short: the two prototypes, 001 and 002, were reminders, with the "small" engine and the reduced weight aircraft, of the "Experimentation Aircraft" by the end of the war.

In a small specification outline of the HA-300 from October 31, 1962, from the "Messerschmitt Study Bureau", the authors confidently declared the beginning of 1963 as the deadline for the aircraft's maiden flight to take place. Otherwise, the remainder of the data and information was quite sparse in this document. The full-view drawing was designed from this paper and shows the aircraft that had its "Roll out" onto the airfield of Heluan at the beginning of 1963.

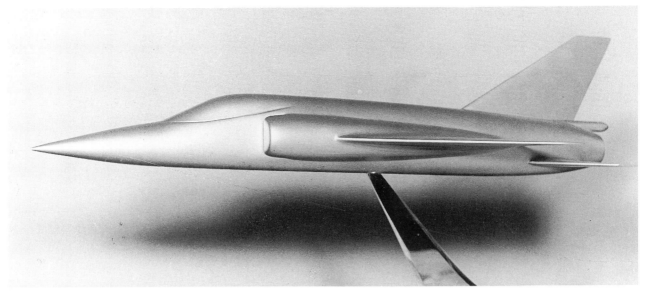

Still, over one year would pass until the aircraft was to perform its maiden flight.

HA-300

Preface

This aircraft was projected on behalf of the Spanish Government with the Hispano Aviacion SA from the German Team of Messerschmitt AG under the direction of Professor Messerschmitt and further developed to a certain point in construction with the support of the Construction Bureau of the H.A.

After the acquisition of the documents from another country, the aircraft was further developed to reach its design maturity.

General Data

The HA-300 is the result of the optimization of the lightest, smallest possible, modern supersonic fighter aircraft with the Mach number reaching two and maintaining sufficient combat power. In the Messerschmitt tradition, the aircraft was built with the lowest possible expenditure to reach the most success. The tradition was thus continued.

A single-jet engine monoplane with triangular wings and horizontal stabilizer in all-metal construction in light alloy;

Hydraulically retractable main landing gear and nose landing gear; landing flap and dive brake;

Sufficient space for various electronic equipment;

Installation possibility of solid firearms, missiles, and jettison loads.

HA-300
Light supersonic combat aircraft

Aircraft data and weights
Crew   1
Power plant   Single-jet engine with afterburner
Static thrust   -
Gross weight dependent
on variant   4200 kg to 5500 kg
Flight characteristics
Maximum Mach
number   2
Operating radius
according to variant   up to 900 km

Apart from all the technical and organizational difficulties, personnel problems also surfaced: who could, who was allowed, or who should take over the flight test prototypes?

Egypt at this time had outstanding Air Force pilots, but not a single training pilot. The German Headquarters desired a test pilot from Europe. And those wanting to partake were unavailable to begin work straight away, as most were busy in involved in long-term test programs.

At the beginning of 1963, a delegation from the Nile stayed with Air Marshall M. Sidky Mahmoud in India and visited the production complexes of the Hindustan Aeronauties (HAL) in Bangalore. The Indians, for their part, continued the talks by dispatching the Indian vice-president Dr. Zahir Hussain to the Near East. A result of this contact was the sending of an HAL test pilot Kapil Bhargava to Heluan. The Indians arrived their in June, 1963, and immediately wanted to take over the test program. After things were straightened out, the first taxi tests began. Difficulties surfaced with the nose wheel.

After these problems were removed, the HA-300.001 was able to taxi in front of President Nassar on the occasion of the Egyptian National Holiday on July 23.

The Indian pilot later proved not to be a an outstanding pilot at all, but rather an experienced technician. He took a detailed look at the aircraft and drew up a list of 19 points which, according to his opinion, were to be improved or even removed. Only then did he have a viewpoint concerning a successful maiden flight. The principal item of criticism concerned the fixed fuel tank, which was a continuous source of problems, and for the Mach 2 fighter, a "simple" but unstable control system and lubrication system, and in the end the lubrication system of the "Orpheus" engine only allowed inverted flight at a maximum of 10 seconds time length.

After a series of discrepancies and misunderstandings the HAL senior pilot, Suranjan, forbid his subordinates from flying the HA-300!

In October 1963, Messerschmitt and his German staff consented to the proposals of the Indians, and, according to Bhargava, "a prolific time of cooperation right up to the end of the project."

German and Egyptian engineers worked on the proposed improvements and modifications of the prototype until the beginning of 1964, then the improved aircraft could retry its taxi attempt with only one year's delay in February, 1964. The long, "comfortable" taxi way of the Heluan airfield and the excellent braking system of the HA-300 permitted the aircraft to accelerate until immediately before lift-off.

On March 7, 1964, it was finally time: the Indian test pilot assumed the maiden flight of the smallest Mach-2 fighter aircraft in the entire world. Two HA-200 aircraft accompanied the fighter on the 12.5 minute flight. For this flight, the landing gear was not retracted. After the landing, the prototype and its pilot were surrounded in a flash by hundreds of enthusiastic workers.

The aircraft's weight for its first flight amounted to 3200 kg; the thrust was limited to 1900 kp.

Bhargava compared the flight characteristics of the HA-300 to those of the "Gnat", which he was most aware of from Bangalore.

The earlier phase of the test program took place within the expansion of the flight speed range both high and low. Many of the problems to follow the HA-300 were a direct result of the new type of design of the HA-300 with its extremely thin wing assembly and control elements. The pilot would climb, always aware of the flutter hazard, and very cautious of the speed, but in the end, it was foreseeable that the flutter of the horizontal stabilizer would take place during high and maximum flight speeds.

With the test flight and the vibration tests on the ground, made all participants were well aware that extensive modifications were to take place on the control system and with the aileron manipulation. In particular, the installation of hydraulic dampers was not to be circumvented. While testing the 001, they also evaluated the various measures and their affects on flight characteristics. During one flight test, they foresaw a new nose configuration without a vertical fin and boundary layer fences in the outer wing area of the prototype.

HA-300.001: during its maiden flight, the undercarriage remained in the extended position. (left)

Drag-chute landing following the successful maiden flight. (below)

156

Flight test of the HA-300.001: notice the boundary layer fences on the wing assembly and the missing vertical stabilizer on the nose. (left)

Photo of the touched-up HA-300.001, for the press release by the Egyptians (above)

The entire design itself was convincing, so it is no surprise that in 1964, the EWR, with its Design 328 for a combat aircraft with vertical take-off and landing characteristics, once again reverted back to the somewhat increased wing assembly geometry of the HA-300. After Messerschmitt had the first results in his hand, he worked on the final production series design for the Egyptian Air Force with the E-300 engine. This engine passed its first test floor run in July of 1963.

Despite the first, problem-laden successes, the opposition won heavy influence against the entire project within the Egyptian leadership. The opposers cited, above all, the high costs and the foreseeable lengthy start-up time as their main arguments.

The German public and the German technical press at this time took slight notice of the goings-on in the Arab country. This favored the Egyptians, who were not particularly interested in extensive publicity for obvious reasons, and neither were they interested in the German Government, who'd rather exchange the "guest workers" on the Nile with tourists in order to spare themselves complications with Israel.

The costs were horrendous to Egypt, and in July 1965, a difference between the construction company and the government over the extreme slightest thing would have split the program. Also, within the Egyptian Air Force, the resistance against their very own, national development program grew visibly, in view of the Israeli strength with its "Mirage"; therefore, an equally strong fighter was required in larger numbers.

The work continued but with less materiel and with corresponding deceleration.

On July 22, 1965, the second prototype completed its maiden flight, and on the very next day, the "Day of the Republic", Bhargava carried out a test flight for Nasser and a list of invited guests. The 002 had, in comparison with the 001, a modified nose wheel, an improved control system, and more extensive electronic equipment. Shortly after commencing the flight test, they installed the reciprocating control to compensate, or battle, the flutter. Similar types would be the horizontal stabilizers of the MiG-29 and MiG-21.

With the weak engine, the HA-300.002 could only reach an altitude of 12,000 meters and a maximum speed of Mach 1.13, wherein the pilot would "assist" the engine somewhat with short dives.

In agreement with the test results in Bedford, the pilot noticed no irregularities or peculiarities whatsoever when breaking the sound barrier. Kapil Bhargava expressed the conviction that, with the intended E-300 engine, the performance parameters would have been able to have been met without a doubt.

By the end of 1965, the four test models of the Brandner engine had attained a total running time of 2000 hours on the test stands. The time for a flight test had finally arrived. Though a combined test flight for a new engine and a new aircraft is a very tricky undertaking, and this hadn't been proposed yet for the HA-300, Brandner's workers re-equipped a Soviet An-12 from the Egyptian Air Force as a flying test-bed. The left, inner turbo-prop engine of the Antonov aircraft was replaced with a E-300 (V9) jet engine, and by June 1966, the flight test was underway.

The well established contacts with India then led, almost simultaneously, to the surrender of the HF-24 "Marut" for testing purposes in July 1966. The aircraft received a built-in "Orpheus" 703 engine system from an E-300 and carried out over 150 flight tests in this configuration, named IBX, by July 1969.

Meanwhile, for the flight testing of the HA-300.002, two Air Force pilots, Zohair Shalaby and Sobhy el Tawil, joined in the testing.

The 002 in front of one of the large engine test beds.
(large photo above)

The Antonov An-12, retooled to test the E-300. (left)

Ferdinand Brandner in front of an An-12 with built-in HA-300-engine E-300. (right)

While on a fighter-bomber raid in his MiG-17 at the beginning of June, 1967, Zohair Shalaby died in the Six-Day War, during which the Israeli Air Force also attacked Heluan.

Professor Messerschmitt hardly stayed in Egypt in 1966/67. As a rule, his employees carried out the work on the HA-300 in Munich. The role in the Federal Republic monopolized the entire middle 1970s and therefore it was no major event that Professor Messerschmitt finally discontinued his activities for Egypt in the summer of 1967.

The former technical director of Focke-Wulf Works, Engineer Kurt Tank, arrived from India to assume the position as director of Messerschmitt's HA-300 development team. He attempted once more to awaken the deteriorating interests of the Egyptians and emphasized the operational capabilities of the HA-300 as a ground attack aircraft, interceptor, and high-altitude fighter able to reach an altitude of 20 km. However, a change in the leadership of the Ministry came in useful for the adversaries, who were convinced of the futility of the program. Additionally, the Soviets, interested in further sales of their MiG aircraft, gained increasing influence.

For the third prototype, Brandner's Engine Factory Number 135 still had two engines available.

High-performance fighter aircraft

for Mach 2

(Scale 1:10)

Design and construction

Professor Messerschmitt

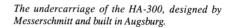

The undercarriage of the HA-300, designed by Messerschmitt and built in Augsburg.

Model (Scale: 1:10) of a "high-performance aircraft built for Mach 2 capabilities" during an air show in Hannover in 1966. The production series configuration of the HA-300 was in question. (below)

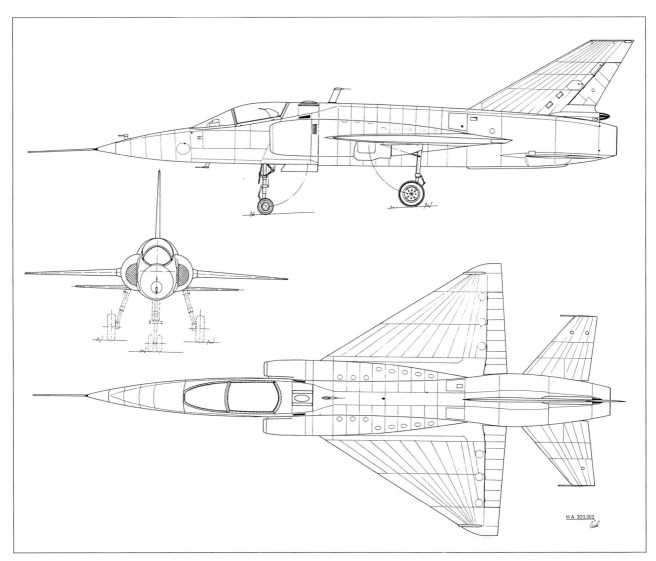

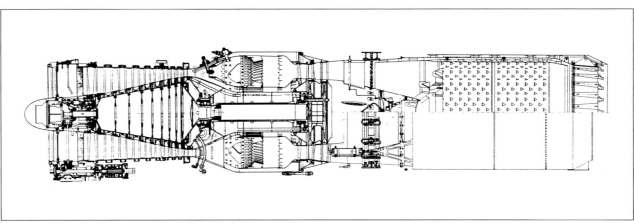

Three-sided view of the HA-300.002. (above)

Cut-away of the E-300. Air flow: 53 kg/second, compressive stress ratio: 1 : 5.7. Temperature in front of the turbine: 1050K, maximum engine diameter: 840 mm, empty weight with afterburner: 860 kg, special consumption: 0.98 kg/kph.

The third prototype was to begin flight testing in the summer of 1969, and was to confirm all performance capabilities. However, they knew they should avoid this.

In May of 1969, the decision came to discontinue the plans after ten years of work. With the exception of the Indian test pilot, all foreign workers had to leave the factories which they had built by the first of June, 1969.

The Egyptian Air Force Technical Headquarters carried out taxiing tests with the 003 in November, 1969, at an acceleration just short of take-off speed. However, technical problems with the engine control forced the relatively inexperienced Egyptians to finally give up.

Over this Land of the Nile, for years there had already been widespread sightings of an aircraft with a silhouette very similar to that of the HA-300 in their skies: In Egypt, just as in India, the contours of the Soviet MiG-21 were well known.

In Factory Numbers 36 and 135 of the industrial complexes of Heluan, native experts repaired light ground attack aircraft called the "Alphajet."

Technical Data: The geometrical data of the HA-300 is taken from Messerschmitt designs and refers to the constructed prototypes.

Three-sided view of the MiG-21 F.

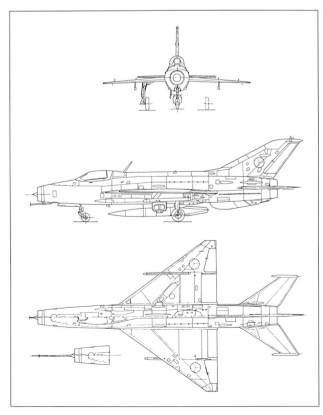

	Heluan HA-300	MiG-21 F13
Primary function	single-seat fighter-interceptor	single-seat fighter aircraft
Crew	1 pilot in pressurized cockpit with ejection seat	1 pilot in pressurized cockpit with ejection seat
Power Plant	1 x EGAO E-300 jet engine with single shaft construction	1 x Tumanski R-11 F-300 jet engine with two-shaft construction
	Static thrust: 3300 kp	Static thrust: 3900 kp
	With afterburner 4800 kp	With afterburner: 4900 kp-5750 kp
	With the HA-300-003, thrust restricted to 3800 kp	
Wing span	5.84 meters	7.15 meters
Wing surface	16.7 m²	23.0 m²
Aspect ratio	2.04	2.22
Contour	bi-convex arched profile	Ts AGI S-12
Profile of the wing	3% in 35% thickness	wing root 5% wing tip 4.2%
Sweep of leading edge	58°	57°
Dihedral angle	-2°	-2°
Aircraft length: without static tube	11.75 m	13.46 m
with static tube	approx. 13 m	15.76 m
Height	3.65 m	4.1 m
Vertical tail leading edge sweep	60°	60°
Horizontal stabilizer wing span	3.62 m	3.55 m
Horizontal stabilizer leading edge sweep	40°	55°
Horizontal stabilizer dihedral angle	0°	0°
Undercarriage	Manufactured by Messerschmitt with Dunlop "Maxaret" brakes	
Wheel base	1.96 m	2.69 m
Track	3.16 m	4.81 m
Nose wheel	2 x 270 x 185 mm	1 x 500 x 180 mm

Main undercarriage	2 x 560 x 180 mm	2 x 660 x 200 mm
Take-off weight HA-300.001/003		3200/3800 kg
Series/normal	approx. 4500 kg	7370 kg
maximum	approx. 5450 kg	8625 kg
Wing load		
normal	269 kg/m²	320 kg/m²
maximum	326 kg/m²	375 kg/m²
Maximum speed	(Mach 2.0) at 12,200 meters	(Mach 2.05) at 12,300-18,500 meters altitude
Climbing speed at sea level	190 m/sec	130-140 m/sec
Initial rate of climb	2.5 min. from 0 to 12,200 meters	3.2min. from 0 to 10,000 meters
Ceiling with external load	18,000 meters	19,500 meters
Operational range	1100-1400 km	1670 km
Armament	2 x 20 mm or 2 x 30 mm cannons. 2 x air-to-air guided rockets	2 x NR 30 mm cannons. 2 x air-to-air guided rockets or 2 x UV-16-57 rocket tanks

This tabled overview shows that an actual light fighter with the smallest possible dimensions was realized with the HA-300. The aircraft, with equal development, could also hold its own, performance-wise, against the competition.

Messerschmitt sought to realize the fighter aircraft having maximum weight, the smallest dimensions, and maximum performance, with the development series 1101, 1106, 1110, 1111, and 1112, and last but not least, the HA-300. This type of aircraft stands atop the wish lists of many strategists of the Air Forces. Again and again they attempt to limit the escalating costs and the climbing weights. Verifying this are studies such as the ALR "Piranha" from a Swiss working group and the SAAB JAS.39 "Gripen", a Swedish design.

Until his death in 1978, Messerschmitt concentrated his efforts on passenger and transport aircraft, the "Rotorjet" which was to have two retractable, counter-rotating rotors over the wing assembly vertical take-off and landing characteristics.

The Aviation and Space Company MBB, which still carries the Messerschmitt name in its firm title today, besides taking part in the development and construction of the successful, multi-jet combat aircraft such as the Panavia "Tornado" or the EFA/JF-90, also partook in the development of a single jet-engine "Experimentation Aircraft." Together with the American firm Rockwell, MBB developed the X-31, which was to become a model for a new generation of air superiority fighters.

Appendix

Engine Variants

P1101 with ram-jet: In the fall of 1944, the possibility to equip the P1101 fighter aircraft, then in development, with a ram-jet engine was evaluated with regard to better construction and higher performance.

The goal was to function in installation defense to simultaneously eliminate the shortfalls of the Messerschmitt Me 163 and to offer advantages such as the extremely simple engine configuration and ample latitude in the type of fuel; practically anything was usable which happened to be liquid and combustible. Dr. Lippisch at the Aeronautical Research Institute in Vienna demonstrated that one could even do without the "liquid": he worked on his designs for ram-jet driven delta-winged aircraft with paraffin-moistened, ground coal.

The disadvantage of this engine was admittedly the high fuel consumption in speed ranges below Mach 2 and the necessity of an auxiliary engine for starting.

At the German Research Institute for Gliders in Ainring near Rosenheim, these were fully developed, along with the engine type named after its inventor Rene Lorin, by Professor Eugen Sänger and his assistant and wife, Dr. Irene Bredt.

By the end of the war at some German aircraft firms (Messerschmitt, Heinkel, Skoda-Kauba), airframes for this engine type (by Professor Sänger) were designed, based on the theoretical work and the practical evaluations. Other firms and institutes, such as Focke-Wulf and the Aeronautical Research Institute in Vienna, worked on configurations based on Lorin's idea.

The use in an inappropriate speed range for a pure ram-jet engine (not to be confused with the "Schmidt-Argus valve"!), thereby causing correspondingly high fuel consumption, would not allow any proposed projects with this engine type to graduate from the drawing-board to the skies.

Immediately following the war, Dr. S‰nger, along with his wife, was employed in an advisory capacity by the French Aviation Ministry. Therefore, he had considerable influence over the development and construction of the Nord Griffon II.

Evaluation of the ram-jet engine under the direction of Professor Sänger (beginning in 1942).

This prototype revolved around a high-speed aircraft with a compound power plant (for starting, a ram-jet engine and for low-speed flight, a jet engine).

In this configuration, at flight speeds in excess of Mach 2, the engine certainly has a future. And not only this type of ram-jet engine found widespread use today, but the well known afterburner as well.

At the discussion in the fall of 1944 about the proposal, the question concerned a P1101 production series whose fuselage was to be enlarged to house the power plant. The Lorin engine consisted mainly of a double-coned sheet steel valve with a fuel injector attachment.

The forward portion of the engine is constructed like a diffuser; through the diffuser the air flow will be decelerated with simultaneous pressure increase to approximately one-sixth of the flight speed. In the joined combustion chamber, the flowing air (with equal pressure) will be heated by the injected fuel — according to the throttling — to well over 1000 degrees C. The warming of the air flow has an increase in volume, and thereby a corresponding rate of speed increase as a result. The air flow is ejected from the jet pipe at a speed just over that of the flight speed.

Right up to the fuel injection mechanism and the regulator for the outlet section, this engine type has absolutely no moving parts. The simplicity of this type of construction and thus the uncomplicated maintenance would certainly have been a distinct advantage in the last phase of the war.

On another page, one reads of the already mentioned high fuel consumption with all its troublesome results such as short flight duration and high take-off weight and the almost totally unevaluated technical and operational concept of the engine.

The take-off thrust of the Lorin engine is equal to zero; for take-off it must therefore use auxiliary equipment. For the P1101 L, eight solid-propellant rockets with 1000 kp thrust each and a combustion time of six seconds were intended. At take-off the rockets would be ignited simultaneously or in succession. The speed attained in this manner was calculated at 120 meters per second (430 km/h); from this point on the Lorin engine pushed further, to include with increasing thrust along with the speed. In addition, the take-off rockets promised, even with retracted flaps, a very short take-off distance.

With the P1101 L, an undercarriage was to be acquired from the 1101 production series configuration. Through the enlarged fuselage and the abandonment of a maneuverability on the ground, this simplified type of landing gear was made possible. Large wheel well doors were avoided and the wheel base of the main undercarriage was deliberately slight in order to make towing of the aircraft on slim routes possible.

As has already been revealed, the support function was to produce an installation defender. The fighter aircraft were to be dispersed throughout the country without having to depend on built-up airfields. Under the direction of a rather small number of permanent staff, the nearest manned airfield was to provide for the crew and low level maintenance.

If a formation of bombers came into view, take-off was to ensue. Because of the longer flight time in comparison to the Me 163 or the Ba-349 "Natter" and similar configurations, the pilot could engage in combat. During the entire time the formation would fly over the German Empire, they were not to have one quiet minute. The attack against smaller formations and single aircraft, for which guidance was especially necessary, was to remain as before the tasking of the fighter formations.

At Dr. Sänger's proposals for application of a ram-jet engine, General Field Marshall Milch from the German Air Ministry provided the following: "With the fullest respect for your achievements, but to build such an aircraft, it is still much too early."

He was absolutely right, although it was already much too late for the Third Reich.

P1101 with ram-jet engine; sideview of the proposed aircraft; Design: H. Redemann

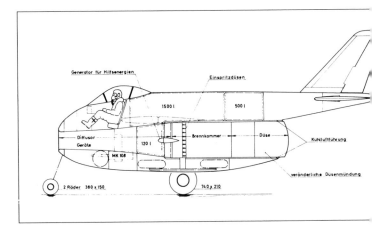

Generator für Hilfsenergien	*= Auxiliary power plant*
Einspritzdüsen	*= Injection valve*
Diffusor	*= diffuser cowling*
Brennkammer	*= combustion chamber*
Düse	*= jet pipe*
Kühluftführung	*= cooling air flow*
2 Röder 380 x 150	*= 2 wheels 380 x 150*
veränderliche Düsenmündung	*= variable nozzle outlet*

Technical Data of the P1101 L Proposal
The dimensions mainly correspond to the P1101 production series configuration. The weight classification also takes the equipment condition of the mass-produced aircraft into consideration.

Weights:

Empty weight:	1900 kg
Crew:	100 kg
Armament:	180 kg
Fuel:	1550 kg
Flight weight:	3730 kg
Lifting aids (8 rockets):	560 kg
Taxiing weight:	4290 kg

Calculated flight performance:
Maximum speed: Mach 0.9
                1080 km/h near ground level
        1000 km/h at altitude of 12,000 meters
Climbing time to 10,000 meters: 3.0 minutes; including take-off time of 1.5 minutes
Ceiling altitude: approximately 15,000 meters
Operating range: (without climbing distance or cruising distance at maximum speed)
 Flight duration
                820 km in 10,000 meters altitude In fuel throttle flight the operational ranges could increase still.
Take-off distance: 150 meters to lift-off (acceleration 2 g)
Landing speed: 165 km/h

Technical Descriptions of the Jumo 004 and HeS 011 Engines

The Jumo 109-004 B: This Junkers engine was the power plant of the first generation of operational turbo-jet aircraft.

The 1092 and 1095 projects are configured with this power plant and since there did not appear to be an end to the problems with the Heinkel HeS 001, Messerschmitt equipped the P1101 with the Jumo 004 as an "alternative engine"; and the first "Experimentation Aircraft" was to be flight tested with the Junkers power plant.

Short description
The eight-stage axial flow compressor possesses eight impellers and nine guide blade rings. In the six separate combustion chambers with baffles, the fuel designated as "J2" is injected and burned over a simplified geared pump ("Barmag" pump).

The turbine is made of an eight-stage impulse turbine, just as with the design by the Firm AEG. The

109-004 B-4 model has concave air-cooled turbine blades. The turbine and compressor shafts are connected to another with serrated teeth.

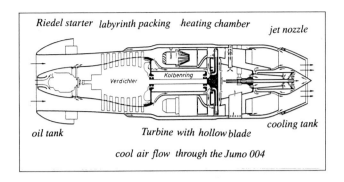

cool air flow through the Jumo 004

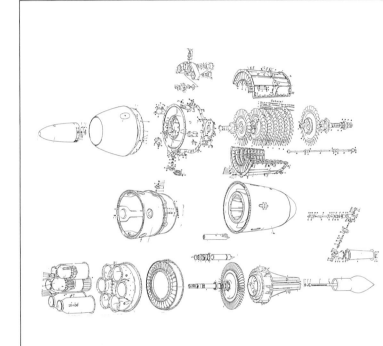

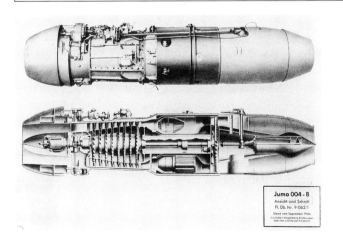

Cross section of the Jumo 004 B

Main components of the Jumo 004 B engine; Jumo 004 B photo and cut-out. (from top to bottom)

The exhaust nozzle is doubled-walled configured for cooling purposes. Control of the engine takes place through a sliding nozzle pintle which either increases (idle) or reduces (full throttle) the exhaust aperture. Control of the engine and the interconnected control of the RPM's takes place with a lever (single-lever manipulation). From the beginning, the jet pipe of the Jumo 004 was equipped for an afterburning power plant.

The engine is connected to the airframe at three points. Inside the cockpit, the pilot monitors:
- the engine RPM's
- the pressure of the combustible gases in the exhaust nozzle
- the gas temperature
- the pressure near the compressor
- the injection pressure and
- the lubricant pressure.

Technical Data:
Design data:

Maximum take-off thrust:	910 kp
Total length:	3864 mm
Maximum diameter:	805 mm
Empty weight, net:	750 kg
RPM's:	8700 min. -1
Specific fuel consumption:	1.42 kg/kp h

Raw Materials:
Turbine: Krupp "Tinidur" (30%) nickel, 15% chromium, 2% titanium, remainder: heavy metal) and later the nickel-free material "cromedur" (13% chromium, 18% manganese, 0.7% vanadium, remainder: heavy metal)
Compressor shapable sheet metal 1010 and combustion chambers: with aluminum protective layer

Compressor:

Total air quantity flow rate:	21.2 kg/second
Air scooping ("Zap Air")	
Cooling jet nozzle	0.5 kg/second
Cooling Turbine	0.7 kg/second
Compression ratio:	3:14
Efficiency in the design point:	78%
Diameter of the primary stage:	545 mm
Diameter of the 8th stage:	582 mm
Combustion chamber:	1 injector nozzle per combustion chamber
Pressure loss	6%
Combustion efficiency	95%

Turbine and jet nozzle:

Turbine housing, inner diameter:	711 mm
Diameter over the impeller tip of the turbine wheel:	698 mm
Blade length:	110 mm
Number of blades:	61
Number of guide blades:	36
Efficiency:	79.5%
Turbine inlet temperature:	775 2 °C
Exit diameter of the jet nozzle:	400 mm
Pressure in the jet nozzle:	1.51 absolute

Starter:
Riedel starter: two-stroke, two cylinder motor with short-duration power output of 10 PS

The Heinkel HeS 109-011: With the jet fighter engine design in response to the RLM call for bids from July, 1944, the emphasis was the same near the end of the war for practically all studies, proposals, and designs from the German aircraft construction builders intended Heinkel-Firth HeS 109-011 A-0 engine (HeS 011 for short).

After they had acquired the well known Hirth engine firm, with this design Heinkel Werke finally dared to attempt to establish German aircraft engine construction;

Heinkel's aircraft designers had already garnered merits through their construction of the first airworthy HeS 3 jet engine tested in the Heinkel He 178 and a whole series of learned experiences and developments from this construction.

The engines preferred by Heinkel with centrifugal construction were not successful against the axial designs from Junkers or BMW, a thought which, years later, had to once again be realized under great pains in British engine construction. Still, the centrifugal engines clearly had their advantages in the earlier phases of jet-propulsion engines; as, for example, in the simple construction, in the manufacturing, and in the operational reliability — a feature the Me 262 engine would have loved to have had.

Be that as it may, those responsible at Heinkel, in any case, went in a new direction with the design of the HeS 011: a major problem of the RLM-preferred axial-flow compressor is the resultant flow separation causing an unfavorable flow ratio. This phenomena, also designated as "pumps", brings on engine "flame-outs" and, in addition, can lead to mechanical damage. In the early years of jet flight, the chance of this type of danger was rather high.

With the construction of the so-called diagonal compressor in the HeS 011, the Heinkel team attempted to bring out the advantages of centrifugal construction type with those of the axial design. Work on this engine was completed in December, 1942, and they were able to begin the construction of yet five prototypes. The first test bench runs took place at the beginning of 1944: during this test the compressor lost its steam after only one hour.

Development was thereby delayed. With a new, unstructured compressor, worked on by the Voith/

Heinkel HeS 109-011 A initial production series engine (from top to bottom).

Heidenheim Firm, the following prototype was equipped. Though the problems with the engine did not find temporary respite. The manufacturing of the diagonal compressor was difficult and costly and only at the beginning of 1945 did the design's thrust reach 1300 kp (12.7 KN).

Following the American invasion, German technicians still manufactured nine HeS 011 A-1 initial production model engines. Engines which were later brought to the USA. In the USA the engines were tested under the leadership of the U.S. Navy beginning in June, 1945, on a test bed in Trenton, New Jersey. The influence of Heinkel construction on American jet engines is not recognizable, which is different in the case of the Junkers on Soviet engines or BMW on the French designs.

Projects for a turboprop engine based on the HeS 011 and further developments for the HeS 011 construction series with a design thrust of 1500 kp and HeS 011 C with a thrust of 1700 kp were still sitting on the drawing board at war's end.

Specification

The compressor of the Heinkel 011 engine is assembled from an axial input stage, a low pressure diagonal stage, and three further axial stages for the high-pressure range. Ergo, the compressor consists of a total of five stages. Today, in the remaining construction type, the combustion chamber connected immediately to the compressor is configured as an annular combustion chamber.

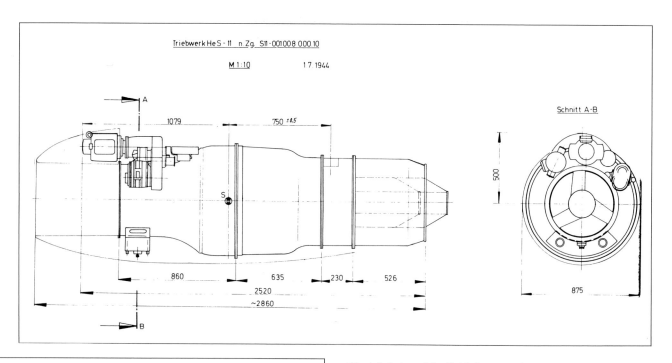

Triebwerk HeS - 11 n. Zg. S11·001008 000.10

M 1:10   1.7.1944

Schnitt A-B

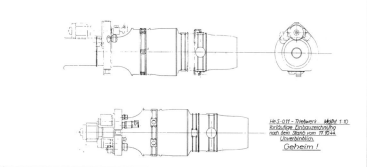

HeS-011 - Triebwerk   Maßst 1:10
Vorläufige Einbauzeichnung
nach dem Stand vom 17.10.44.
Unverbindlich.
Geheim !

"Early" design of the Heinkel engine: the outer shape is sharply varied from the later designs. (July 1, 1944)
Only three months later, the Heinkel engine is shaped based on the rebuilt compressor. (This according to the Messerschmitt diagram from October 17, 1944, and Heinkel design 109-011/1501. 10 - (left)

(From inside lower-left sketch)
HeS-011 engine: scale 1 : 10.
Preliminary diagram according to status from October 17, 1944.
Secret!

On its front side there are duplex nozzles which inject the fuel supplied at an angle of 80 degrees into the combustion chamber.

Ignition takes place with the usual spark plug. To support the combustion process there are small holes in the combustion chamber walls. In the chamber itself there are three blades. There perform the following:

1st blade: Deflecting the combustion gas into the interior of chambers (up to 90°)
2nd blade: Delivery and swirl with additional air to promote combustion.
3rd blade: Directing the secondary air into the interior of the combustion chamber: creation and preservation of an even combustion temperature.

The extraordinarily fine working combustion takes place through this complicated system. The exhaust gas powers the turbine wheels which are equipped with air-cooled hollow blades.

The cooling air required for the hollow blades is taken from behind the third axial stage and led to the blades. The combustion gas is directed to the open air. Just as with the remaining German turbojets, thrust control takes place over a nozzle bulb. Thereby, operating the HeS 011 is possible with "single-lever manipulation." All auxiliary devices and control mechanisms are attached to the forward compressor housing, to include the primary element to monitor the engine.

Technical Data: Construction Data:

Maximum take-off thrust:	1300 kp
Total length:	3450 mm
(with air duct tube and nozzle pintle in the full throttle position)	
Engine length:	2640 mm
(Nozzle pintle in the idle position, without starter)	
Maximum diameter:	830 mm
Empty weight, net:	770 kg

Engine, to include starting
fuel and lubricants: 845 kg
Total engine system:
(fuel tank in airframe)   921 kg
RPM's   11 00 min -1

Compressor:
Total air quantity: 30 kg/second
(at 0 km altitude, V = 0 km/h)
Compression ratio: 4

Combustion chamber:
Annular combustion chamber with 16 injector nozzles;
For starting: 2 atomization jets and
    2 ignition plugs
Fuel pressure: 40 kp/cm^2
Turbine and jet nozzle:
Number of stages: 2 with air-cooled blades
Exhaust temperature: maximum 550° C
    (temporarily 800°C when starting)
Starter: Two-cycle gasoline motor

Flight speed, thrust, and specific fuel consumption during full-throttle operation:

	Speed (km/h)	Thrust (kp)	Specific fuel consumption (kg/kphour)
Alt. 0	0	1300	1.32
	300	1115	1.44
	600	1020	1.60
	900	1040	1.78
Alt. 9 km	0	605	1.12
	300	555	1.26
	600	525	1.38
	900	550	1.48

Equipment/Armament

Probable equipment for a Luftwaffe fighter aircraft in 1945/1946.

Standard equipment: The fighter-interceptor can have a very simplified standard equipment (for example: fighter control of heading and FuG 125 communications package); otherwise, the standard equipment encompasses the following:

- electronic fighter control of heading
- on-board camera
- EZ 42 sighting mechanism for fixed weaponry; this revolves around a programmable sighting assemblage for automatic determination of amount of lead information. Through testing, approximately 30% more targets were hit in comparison to the Revi 16 B despite the greater firing distance.

- FuG 15 remote controllable VHF communications equipment from Lorenz, with instrument unit, transducer and antenna matching unit inside the fuselage aft and an AFN 2 indicator for radio navigation inside the control panel (instrument flight panel).

- FuG 25a communications package from a ground station IFF interrogation system; recognition from aircraft to aircraft is the next step. The system can be used for fighter guidance.

- FuG 125a "Hermine" navigation and direction

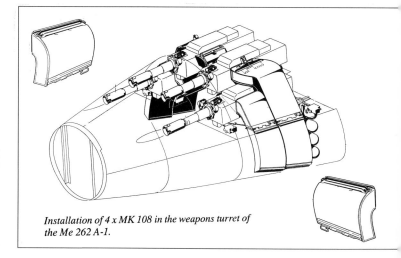

Installation of 4 x MK 108 in the weapons turret of the Me 262 A-1.

finding equipment which guides the fighter aircraft over a beam to its air base. This "radio landing equipment" possesses a remote controllable receiver EBL 3 F inside the rear of the fuselage. With the FuG 120 K attachment, the radio beacon "Bernhard" can be received. The system is equipped especially on one-seater aircraft.

- MK 108 Machine gun; it is planned to be installed as standard equipment as four-two and two as auxiliary cannons. The probable four MK 108's, with 100 rounds of incendiary bombs each, are grouped beneath the cockpit surrounding the air flow valve in the forward fuselage section of the P1101.

With this weapon, made of Rhein metal-Borsig, the theme is the compact, short cannon with a low initial speed. It produces the sound of a compressed air hammer when firing. For engagements against ground targets the MK 108 is unsuitable but against enemy aircraft it is extremely strong. Oftentimes very few hits against a four-engine bomber are required to bring it down.

It operates electro-pneumatically. 8 x 21 compressed-air cylinders are required and are grouped into one unit. Firing of the cannons takes place through manipulation of a knob on the joy stick.

Technical data:
Rhein metal-Borsig MK 108

Caliber	30 mm
Length	1057 mm
Weight	58 kg
Projectile weight	0.33 kg
Cartridge weight	0.48 kg
Muzzle velocity	approx. 530 m/second
Firing velocity	approx. 660 min -1

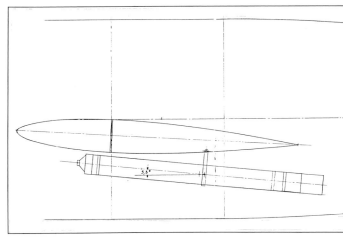

Auxiliary equipment

According to an aircraft's primary function and tactical tasking, the auxiliary equipment may be combined and swapped.

The list doesn't lay claim to completeness since other types of equipment may be included at any time.

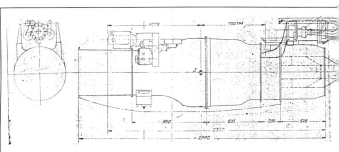

1: Expendable fuel tank under the fuselage for 500 liters of fuel.

2: Droppable rocket boosters for take-off if the aircraft is overloaded. Perhaps the used Rhein metal-Borsig Ri-502's, already on the Me 262, could come in handy in this case. Most of the rockets are in pairs and attached to the underside of the fuselage. They have a thrust of 500 kp that can easily be maintained for the relatively short span of 6 seconds during take-off.

3: Jet propulsion engine with supplementary thrust (built-in, permanent (not jettisonable) rocket engine for the interceptor variant).

This engine enables the fighter-interceptor (the "Homeland Defender") to attain its operational altitude in the shortest possible time. This configuration combines the advantages of the rocket fighter with those of the turbo-jet fighter. Before war's end, the Messerschmitt Firm began evaluation of such an outfitted Me 262 in Lechfeld near Augsburg. Following the war, this idea was continued, in great detail, by the victorious powers. However, the development and production series introduction were futile next to the ever-maturing air defense rocketry, and therefore never completed.

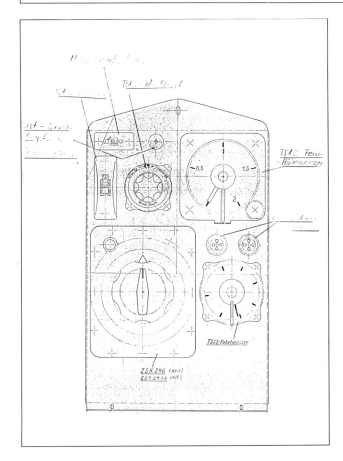

4: Radars for night- and all-weather operation
- FuG 350 Zc "Naxos" with an EA 350 Zb receiver antenna in the fuselage nose and an SG 350 Zc display unit built in to the pilot's instrument panel.

This equipment aids in detecting the opponent and reacts to the highly developed English radar H2S (German designator: Rotterdam Radar). The FuG 350 is a passive radar without a transmitter or keying mechanism; ergo, it cannot be located or even detected by the enemy.

Soviet fighter aircraft from the 1950s (MiG-19, for example) were, in great numbers, with a further developed "Naxos", which reacted well to the modified H2S MK 9A radar used by British bombers.

- FuG 218 "Neptun V" with its Siemens point antenna attached to the fuselage nose and an SG 218 display unit inside the instrument panel. The control unit for the FuG 218 and the toggle switch for the FuG 350 are built in to the right instrument panel. The FuG 218 is an airborne radar for active target seeking and is used for single-seat aircraft. The danger of being detected or of being jammed by the enemy is very great by the highly developed British radar technology.

Equipping this FuG 218 configuration as a "rear warning radar" is possible.

Armament

5: Fixed armament with 4 x MK 108.

Soviet fighter aircraft from the 1950s (MiG-19, for example) were, in great numbers, with a further developed "Naxos", which reacted well to the modified H2S MK 9A radar used by British bombers.

- FuG 218 "Neptun V" with its Siemens point antenna attached to the fuselage nose and an SG 218 display unit inside the instrument panel. The control

unit for the FuG 218 and the toggle switch for the FuG 350 are built in to the right instrument panel. The FuG 218 is an airborne radar for active target seeking and is used for single-seat aircraft. The danger of being detected or of being jammed by the enemy is very great by the highly developed British radar technology.

Equipping this FuG 218 configuration as a "rear warning radar" is possible.

Armament

5: Fixed armament with 4 x MK 108.

6: Fighter bomber.

For fighter-bomber operation, a 503 A-1 bomb release on the fuselage side is attached with a drop bomb totaling 500 kg (for example: 1 x SC 500/SD 500/AB 500-1 or 2 x SC 250/SD 250/AB 250-2). The release bombs are dropped with the aid of a TSA-D2 sighting mechanism. This "release and ejection system" (TSA - "Tiefwurf- und Schleuderanlage") from Zeiss has already been tested in place of the Lofte 7 D with an Me 262A-2. This system has an automatic seeker which is equipped with a control computer and a gyroscope. Together with the reflector sight, it enables rather exact targeting even in bad weather. Only the wind speed must be input on the system; airspeed, altitude, and angle of descent are continuously and automatically updated.

For approximately 20 seconds, the pilot flies a direct route and he can approach the target directly and also detect it in his reflector sight. As soon as a signal is heard in the headphones and the fighter-bomber finds itself in release range, the pilot depresses the bomb release knob and then immediately pulls the aircraft into a steep climb. The weapon is released by the computer only when the aircraft reaches its drop

point. The bomb is "flung" on a parabolic flight route towards its target. By this time the fighter-bomber is a great distance from the striking point.

7: Unguided air-to-air missile R4M "Orkan."

With this prototype for air-to-air rocketry the Luftwaffe achieved its last great success.

Launching takes place through use of salvos from a 12-round grate bar under the wing. The pilot has over 24 rockets available and he fires as salvos of four at timed intervals of 0.07 seconds for each 6 rockets.

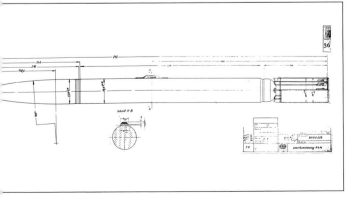

12 x R4M in the grate bar under the wing assembly of a Me-262. The configuration on a single-jet engine fighter would have been similar. (left)

Unguided air-to-air rocket R4M with folding fins - a deadly threat for allied bombers. (below)

Technical data:

Total weight:	approximately 4 kg
Weight of explosives:	0.520 kg
Propellant charge:	0.815 kg
Maximum thrust:	245 kp
Combustion time:	.75 seconds
Maximum speed:	525 m/second
Operating range:	1500 m

In the last years of the war, the WKL (Waffenkonstruction-Luft)-Gruppe of the Rheinmetall-Borsig Firm in Berlin-Marienfelde developed the wing stabilized rocket R 100 BS; that is, loading with (400 rounds) incendiary fragment bombs.

The conversion kit for the series encompasses two AG-140 launchers which are attached to dropable carriers. The launchers consist mainly of three 1800 mm long rails onto which the rocket propelled vehicle with a caliber of 210 mm is attached and from which the rocket is launched and guided. (Later the rails were shortened to 1200 mm and for operational designs, the length of the AG 140 rails were supposed to amount to 800 mm). The launch is produced using the "Oberon" method with the EZ-42, and using the "Oberon" clock and the EG 3 "Elfe 3" release mechanism in connection with the FuG 218 airborne radar or with the FuG 248 range finder "Eule" from Telefunken. The FuG 248, also designated as a "fighter launcher", possesses a funnel-shaped antenna built mostly into the wing and it provides the distance of the enemy via "Elfe" in the EZ 42 sighting system. The operating range of the "Eule" system amounts to approximately 2000 meters and thereby coincides with the operating range of the R 100 BS.

The favorable launch distance is approximately 600 meters distance from the opponent. An explosive charge expels BS munitions, located in the forward section of the rocket, approximately 80 meters in front of the target. For enemy bombers, this munitions forms a deadly dispersion effect with a diameter of up to 100 meters.

Technical data for the R 100 BS:

Caliber:	210 mm
Maximum diameter:	282 mm
Length:	1840 mm
Span of the stabilizer:	320 mm
Weight:	110 kg
Maximum speed:	450 m/second (1610 km/h)
Operating range:	2000 m

9: Wire-guided rocket Ruhrstahl 8-344 (X-4)

The air-to-air guided rocket X-4 is based on the work of Dr. Max Kramer from the DVL. It was built in mass production from the Ruhrstahl Firm AG in Brackwede near Bielefeld. Also during the Third Reich they developed, from the X-4, the wire-guided anti-tank missile X-7 "Rotküppchen", which formed the basis for the successful post-war development Bölkow Bo 810 "COBRA." The long-range X-4 air-to-air missile was a wire-guided missile, launched, as a rule, from an under wing station (ETC 70).

The total for the P1101 was to be four missiles.

Control of the missile is provided via two isolated, unwinding, 5500 meter long and 0.22 mm thick wires. During flight nearing the sound barrier the rockets are rotated approximately 60 min. -1 at the longitudinal axis.

During launch the fighter is located outside enemy defenses and the flight route of the projectile cannot be interfered with by the enemy, thanks to the wire guidance. The pilot provides the guidance input by using a small joy stick in the cockpit to the wire-guidance system FuG 520 "Düsseldorf." To increase the possibility of a direct hit, the missile is able to be equipped with the acoustical seeker "Dogge" and the "Meise" proximity fuse, which reacts to the propeller noise.

The illustration depicts the X-4 without "Dogge" seeker and "Meise" proximity fuse.

In 1944 the weapon was being evaluated:

Production of the guided rocket in mass quantities was hindered by the allies bombing raids which neutralized the manufacturing of the BMW 109-548 power plant.

By the end of the war no effective use of this superior weapon was yielded.

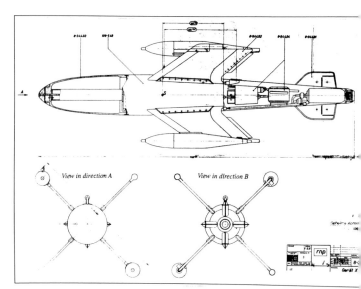

Wire guided missile X-4: an example of this weapon lies in the Deutsches Museum today.

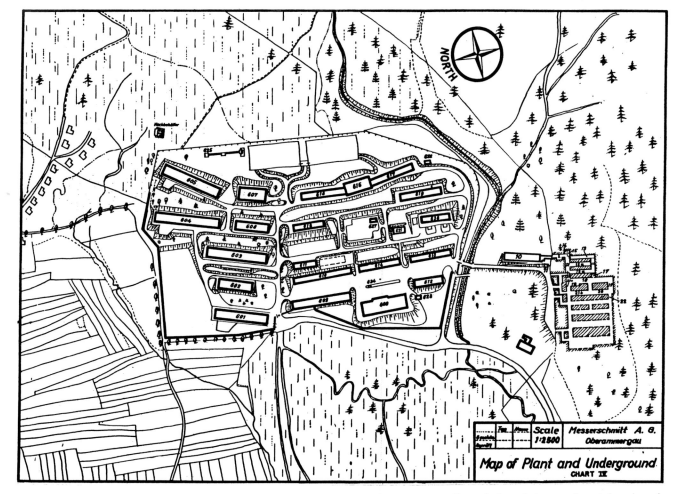

Ground plan of the complex at the edge of Oberammergau. The underground portions of the complex ("Messerschmitt Tunnel") are shaded.

Technical data

Length (without "Dogge" seeker and proximity fuse):	1702 mm
Length (with fuse):	2001 mm
Diameter of missile:	222 mm
Maximum span of the crossed wing with wire spool:	approx. 780 mm
Span with stabilizer:	275 mm
Length of forward section of missile (warhead):	450 mm
Total weight:	approx. 60.5 kg
Weight of explosive charge:	approx. 20.0 kg
Operating range:	approx. 5000 meters
Maximum speed:	248 m/second (900 km/h)
Power plant:	
Type:	Liquid propelled rocket engine BMW 109-548
Thrust:	Maximum 140 kp
Combustion time:	approx. 22 seconds

The Upper-Bavarian Research Institute in Oberammergau

Since weighty bombing raids on the Messerschmitt Main Factory in Augsburg-Haunstetten were feared, in the fall of 1943 the Project Bureau, the Development Bureau, and the Construction Bureau began resettling in Oberammergau in an empty mountain soldier lodging facility ("Communications Barracks") on the eastern edge. The technical leadership found room and board in the hotel "Wittelsbacher Hof." Messerschmitt himself did not remain continuously in Oberammergau since he also maintained an office in Augsburg.

Moving the prototype construction began in the fall of 1944 and was completed sometime in June of the following year. Before assembly of the P1101 began in Hangar 615 in the fall of 1944, the workers of the test construction had equipped two Me 262's with an additional rocket engine. ("Homeland Defender").

Aerial photograph of the Research Institute near the end of the war. (above)

Underground portion of the complex while under construction. (left)

The offices and hangars of the Messerschmitt AG were located at the edge of the ideal mountain retreat, well known for its Passion Play.

Immediately behind the stretching building, the walled landscape rose rapidly. Inside this mountain slope, underground hangars and tunnels were built: the construction and development of this part of the complex was not yet completed by war's end. In the full-view drawing, the underground manufacturing area and supply areas are shaded. Despite the incomplete construction work, the tunnels were already being used.

In the area around Oberammergau, such as, for example, in Garmisch-Partenkirchen (material testing, propeller development) and in Linderhof ("Enzian") there were additional offices of the "Oberbayerische Forschungsanstalt" complex. This complex, in order to describe it with a modern term, was a "high-tech-nology center", and was never detected by allied aerial reconnaissance and was therefore spared from air attacks.

Today the former Messerschmitt buildings are now the accommodations for a military school for NATO.

Tabular Summary of Development

This summation, and mainly the designations, lays no claim to completeness; the authors are thankful for any information supplied or corrections.

Date 1921	Feasible, constructive proposal from Guillaume/France for a gas turbine aircraft with multiple stage turbo fans, annular combustion chamber, and multi-stage turbine.
1929	Proposals regarding a jet power plant for aircraft from Frank Whittle is rejected by the British Ministry of Aviation.
1930	Frank Whittle receives a patent for a turbojet aircraft. Independent from this, a physics student in Germany named v. Ohain works on realization of this type of engine.
1933	Announcement from A. Busemann and O. Walchner: "Profile Characteristics in Supersonic Speeds."
1934	In his articles "For External Ballistics of Rocket Aircraft" and "Rocket Aircraft in Active Air Defense", Professor Sänger illustrated, with support from the works of Prandtl, Hohmann, Busemann, and Ackeret, among others, the path to a supersonic interceptor aircraft. Some of the problems inherent in supersonic flight became known and were described.
1935	The later Heinkel employee, Hans Pabst von Ohain, receives the patent for the jet-propulsion engine with radial flow compressor, annular combustion chamber, and a radial turbine.
1 March 1935	Official "de-masking" of the German Luftwaffe.
28 May 1935	The prototype of the Bf 109D-IABI makes its maiden flight with a Rolls-Royce "Kestrel" engine.
30 Sep-6 Oct 1935	"Le alte velocité in aviatione" XIII. Reale Accademia d'Italia Fondatione Allesandro Volta, the so-called Volta-Congress in Rome. Adolph Busemann (Aeronautical Research Institute LFA) presents the lecture "Aerodynamic Lift in Supersonic Flight Speeds", which represented the first mentioning of swept wing characteristics for reduction in resistance in high-speed flight.
26 Nov 1937	At the third Scientific Conference of the German Academy for Aviation Research, Messerschmitt presents a lecture regarding the problems of high-speed flight and thereby plainly calls for increased research into sonic speed. Until the end of the war, Professor Messerschmitt maintained a leadership position in German aerodynamic research.
March 1938	Walter Rether, arriving from Arado, becomes Director of Construction at Messerschmitt.
11 July 1938	Willy Messerschmitt becomes General Director of the current "Messerschmitt AG."
6 Sep 1938	Along with Heinkel, Porsche, and Todt, Messerschmitt receives the National Prize for Science and Art.
October 1938	The first, single-engine design for the Me 262, evaluation of all possible accommodations for the engine. These early studies were the first step, albeit dissimilar, to the P1101.
2 Jan 1939	Alexander Lippisch, along with his staff, is integrated into the Messerschmitt AG.

4 Jan 1939	The RLM brings on the "temporary technical guidelines for high-speed fighter aircraft with jet-propulsion engine", whereby 900 km/h for maximum speed of a jet-propelled fighter is already being proposed.
5 March 1939	Engineer Lusser, Director of the Project Bureau at Messerschmitt, departs in favor of Heinkel, and Engineer Voigt becomes his successor.
During 1939	Professor Albert Betz (AVA Göttingen) begins systematic work on swept wing research. The testing is conducted by Ledwieg and Strass.
26 April 1939	Fritz Wendel flies the Me 209 to an absolute world speed record (755.11 km/h). The end of propeller-driven, piston engine fighter is clearly evident.
27 Aug 1939	The first turbo-jet driven aircraft in the world, the Heinkel He 178 takes-off for its maiden flight and introduces the "jet age."
9 Sep 1939	Betz and Busemann receive secret patent Number 732/42 "Near Sonic Aircraft Speeds."
Beginning of 1940	Testing the swept wing at the AVA Göttingen by order of Messerschmitt AG. The tests revolved around a three-component measurement on 16 different swept wing models in speed ranges from Mach 0.6-0.9.
13 Aug 1940	Design of the parasite fighter P1073 B. Project P1073 remote bomber as launching aircraft for three parasite fighters. Researcher: Hornung.
3 and 26/27 Sep 1940	Conferences of the Committee for General Air Flow Research; theme: "Maximum Speed" from the Lilienthal-Association for Aeronautical Research (LGL) in Göttingen and Braunschweig; explanation of swept wing research work and its results on the aviation industry. Busemann advises on the swept wing in various articles, such as, for example, Yearbook of German Aviation 1941 "Aerodynamic Wing Construction. The influence of Mach Numbers."

1941	Patent application from Alexander Lippisch "Swept Wing With Rotating Outer Wing Sections."
April 1941	Me 262 proposal with 35 degree wing sweep.
June 1941	Wind tunnel tests with this high-speed wing.
2 Oct 1941	H Dittmar flies 1004 km/h in an Me 262 in Peenemünde, Germany.
18 July 1942	Maiden flight of the Me 262 V3 with pure jet propulsion (Junkers T1); an aircraft designed without taking the swept wing effect into consideration; the low wing sweep was the cause of a displacement of the center of gravity which gave rise to the installation of the heavy Junkers engine. Planned further development of the Me 262 swept wings were to be carried out on swept wings with a sweep of 45 degrees. Beginning of 1943 Initial working of the firms' wind tunnel in Augsburg-Haunstetten.
End of May until beginning of July 1943	Project P1092 for a multi-purpose combat aircraft and a single-jet engine fighter. Project work considered, among other things, a comparison with on-going aircraft programs (Me 262). Researcher: Hornung
July 1943	Messerschmitt employee Engineer H.J. Puffert was named as "Specialist for Swept Wings in the Wind Tunnel Special Committee."
Autumn 1943	Beginning of the relocation of the Project Department and Construction Department from Augsburg to Oberammergau. Test construction began in March of 1944.

Oct-Nov 1943 Project P1095 for a small and feasible single-jet engine fighter aircraft made from available main components. Researcher: Seitz

1 March 1944 Formation of the "Fighter Staff."

21 Mar 1944 Junkers employees Hertel, Frenzel, and Hempel apply for a patent with the title: "Low resistance construction of high-speed aircraft, and also such phenomena outside the scope of aircraft displacing bodies." The content of this patent specification anticipated the area rule concept from Whitcomb. Some German aircraft projects (among others, the Me P1110, He P1073, and Focke-Wulf 1000/1000/1000 were already designed according to this "find" at the end of the war.

First-half of 1944 Combat aircraft and bomber projects under the designation P1101. Designs under further development of the P1100 project with two, three, and four Heinkel HeS 011 engines. This concerned studies with wing sweeps of 35/40/45 degrees, with main rotor for vertical take-off and landing, with crescent wing, and with variable geometric wings.

First-half of 1944 The pressure from the allied bombing raids and the appearance of the American long- range escort fighters led to the down grading of the German bomber program and to the establishment of the "Emergency Fighter Program."

20 June 1944 In the RLM there was consideration surrounding arrangement of such a program.

25 June 1944 By telephone, Hitler gave Speer and Saur the mission to carry out introduction of a large-scale fighter program. The goal: manufacturing of 5000 fighter aircraft per month.

15 July 1944 Under Program Number 226/2, the totally unorganized Technical Office of the RLM announced, in respect to the fighter program, a call for bids for a single-jet engine fighter aircraft. Focke-Wulf, Heinkel, and Messerschmitt received the first develop ment contract and, somewhat later, Junkers, Blohm and Voss were included. All the firms provided designs with swept wing assemblies.

24 July 1944 First fighter aircraft design with project number P1101 following the competitive tendering from July 15, 1944. The design still does not possess any similarities with the later configured design.

22 Aug 1944 Design study from Thieme for the single-jet engine fighter with high-set engine and V-shaped horizontal stabilizer. This design illustrates similarities to the later configured He 162.

29 and 30 Aug 1944 First design, as desired by the RLM, for the P1101: expert on hand was the Director of Project Group 1 in the Project Bureau, Engineer Hans Hornung. "As desired by the RLM" means: 1 full-view drawing in scale of 1 : 50 1 long-view sketch in scale of 1 : 10 1 set of guidelines which must contain: a) Maximum speed near ground level b) Maximum speed at 3000 m altitude c) Maximum speed at 5000 m altitude d) Land ing speed e) Climbing speed f) Service ceiling g) Flight range 1 speed configuration for primary role as: a) high-speed fighter aircraft, and b) "Homeland Defender" (interceptor) 1 specification sheet 1 Design list (from the "Guidelines for Fighter Aircraft with Jet-Propulsion Engine" from January 4, 1939).

8 Sep 1944 Above all, the delays in development of the HeS 011 and the further intensification of the raw material situation led the RLM to a new call for bids in respect to the 226/2 Program.

Apparently, in the RLM the knowledge had been achieved that the technically ambitious bid set up in July, 1944 was not to be realized in the foreseeable future. Now they called for: a simple and, above all, quick-to-be-produced fighter aircraft with a BMW 003 jet-propulsion engine: victor of this bidding procedure is the Heinkel He 162.

10 Sep 1944   At the instigation of the RLM, the first comparison of the jet-propelled fighter design takes place in Oberammergau. Focke-Wulf takes part with the "Flitzer" Project, Heinkel with a somewhat enlarged and configured with swept wings He 162, and Messerschmitt with the P1101 design from August 30, 1944. Messerschmitt, Certificate Number 1: Comparison of the jet-propelled fighter designs.

13 Sep 1944   Publication of the Messerschmitt construction guidelines for the coming production series aircraft, P1101. Decision for construction of an "Experimentation Aircraft."

3 Oct 1944   New design for the P1101: out of this, wind-tunnel model 1 : 10 for aerodynamic testing at the DVL.

28 Oct 1944   Design of the wind tunnel model. Testing a P1101 configuration with Lorin-valve; engine based on the work of Professor Sånger at the DFS (German Research Institute for Gliding Flight) in Ainring.

2 Nov 1944   Equipment plans for the P1101 production series aircraft.

8 Nov 1944   Project hand over 1101. Hand over of data and documents from conception to construction. Documents from Hornung and Voigt are signed.

13 Nov 1944   Full-view drawing: P1101 production series aircraft.

End of Oct, Beginning of Nov 1944   Problems with the 1101 design: considering an improved proposed design.

15 Nov 1944   Attempt to test the loss at intake (Evaluation Report from November 23, 1944).

Middle Nov   "Miniature Fighter" competitive bidding: even the data from the September call for bids doesn't suffice in view of the situation becoming rather desperate. The RLM now calls for the simplest fighter with a pulse engine (Argus 109-014, the V1 engine), which was supposedly quicker to produce than the Heinkel He 162; that is to say, shortly before the end of the war the (aviation) industry struggled with three different calls for bids for a fighter- interceptor! This "waste" was the result of an incomprehensibly late reaction by the leadership of the air war to the Anglo-American threats and the general situation in regards to the war and raw material. For further aviation technology development, only the bidding competition from July, 1944, can be designated as relevant.

15 Nov 1944   At the instigation of Messerschmitt, evaluations begin at the AVA Göttingen for the high-set horizontal stabilizer. Work was completed on February 24, 1945.

4 Dec 1944   Initial construction work for the "Experimentation Aircraft" — selection of construction materials for the P1101.

4 Dec 1944   First design for the P1106, design XVIII/144 from Büchler.

12 Dec 1944   Full-view sketch of the XVIII/151 for the P1106: design with T-shaped horizontal stabilizer.

14 Dec 1944   From the P1106 proposals for a supersonic rocket aircraft and for a P1106 design with increased operating range (design XVIII/154).

14 Dec 1944   Construction specification: "Single-seat, single-jet engine fighter P1101" (Production series aircraft). Afterwards Decision on RLM presentation in favor of the P1106

15 Dec 1944   Discussion in the Headquarters of the Luftwaffe concerning a new single-jet engine fighter. Messerschmitt decides to produce the P1106 following the performance comparison.

19-21 Dec 1944 Performance and flight characteristics comparison of the designs submitted at the DVL in Berlin, Adlershof under the direction of Professor Quick. Experiment to carry out a computation adjustment. The participants agree to general acceptance in regards to maximum speed and Mach influence and establish the revision of the designs and resubmission for a conference in January, 1945. The 1106 project receives an unfavorable assessment in its present form.

8 Jan 1945   Design: V-shaped horizontal stabilizer for P1101.

9 Jan 1945   DVL report from the December conference: "Adjustment of the fighter project with the He 11 jet engine" (A.W. Quick and Höhler). In the meantime, the Messerschmitt Bureau developed the P1110 design from the 1106 project (with circular-shaped air intake, V-shaped horizontal stabilizer, and the wing assembly of the 1106.

11 Jan 1945   Design XVIII/159 for the P1106; the project was submitted in this form to the RLM and the DVL. New recommendations from the General Staff for the single-jet engine fighter: 2 hours flight time with 100% thrust and armament consisting of 4 x 30 mm weapons. The war diary is annotated thus: "the recommendations will not be able to be fulfilled."

12 Jan 1945   First P1110 design with circular air intakes and V-shaped tail unit for the DVL submission.

12-15 Jan 1945 Further conference at the DVL in Berlin, Adlershof, with firms taking part in the development of the new, single-jet engine fighter aircraft. Messerschmitt submits the P1110 project in its original form, derived from the P1110 project. The "Special Commission Day Fighter" will turn in the construction specifications for the new jet fighter design. Discussion about the designs and final establishment of the guidelines for requirements of the basis of the performance calculation. The project from Blohm and Voss, Junkers, and the new Focke-Wulf project left the best impression behind. Messerschmitt reworks the P1110 project and develops the P1111 as a tailless alternative design.

18 Jan 1945   Prototype conference for the P1101 Number 1: status of the test-bed aircraft- manufacturing, deadline for flight ready testing was established as the 15th of March, 1945. (Because of the lack of an engine).

29 Jan 1945   Conference for P1101 prototype Number 2: Second area rule of the test aircraft is ready in Leonberg.

End of Jan   At the EHK the Junkers EF-128 project and the Messerschmitt 1110 project are proposed for further development (not for production series manufacturing!).

1 Feb 1945   Hügelschäffer takes over research and project control for the P1101 from Seifert.

2 Feb 1945   Design XVIII/165, fundamentally improved P1110 proposal. Being reworked: new wing assembly and intake.

12 Feb 1945   Study: P1110 "Duck." Researcher: Büchler. Work on the P1111

13 Feb 1945   Conference by technical leadership regarding the 1101.

15 Feb 1945   Disagreements at the RLM: HDL Saur sharply turns against Chief of the TLR Diesing and forbids several firms from accepting development orders from the Chief of the TLR-F1.E (Knemeyer).

19 Feb 1945   Meeting with Professor Messerschmitt concerning the 1101.

20 Feb 1945   P1101 production series design with forward-retractable undercarriage.

21 Feb 1945   Full-view drawing of the P1106 III/725. (The design is no longer a factor in view of the almost completed P1101 design).

21 Feb 1945   Prototype conference number 3. 2nd area rule fulfilled in Oberammergau. The engine or the 1st prototype is still not available. P1101 brought up to undergo component testing for the P1110; the modifications mainly revolve around the wing assembly and tail unit.

22 Feb 1945   Full-view drawing III/726 of the P1101 production series aircraft.

22 Feb 1945   P1110 full-view drawing following retooling with improved undercarriage; proposal presented in its form from February 22.

24 Feb 1945   XVIII/168 full-view drawing for P1111.

25 Feb 1945   P1111 revised to the P1112 night fighter proposal.

26 Feb 1945   Technical Report TB 139/45 from Hornung/Voigt: "Turbo-jet fighter with and HeS 011 engine" provides for preparation for conference from 27 and 28 February.

27-28 Feb 1945 From the "Final Conference" solicited by the RLM at the EHK concerning the new turbo-jet fighter: Messerschmitt presents the P1101, P1110, and P1111 projects. The following agreements were established: Focke-Wulf Project Number 279 (Design 2) for "immediate resolution." (Rejected by the TLR-F1.E). In connection with this Messerschmitt reworks the P1111 project to the P1112.

28 Feb 1945   Design III/727: Wing assembly "A" for the "Experimentation Aircraft."

Beginning-Middle
March 1945   Besides Messerschmitt, whose P1101 is presently in the construction stage, Focke-Wulf, Blohm and Voss, and Junkers fully intend to build prototype aircraft: there are orders at the RLM. TLR F1.-E. calls for the following projects to be developed: 1. Junkers, Blohm and Voss and a modified Focke-Wulf project; 2. Messerschmitt optimal solution, and in order to increase the confusion, they propose to the EHK to link the project of the Henschel Firm as the second result of the optimal solution with the introduction of a ram-jet fighter using solid fuel by Lippisch and Heinkel.

3 Mar 1945   Design P1112/S2

6 Mar 1945   Design: Tail unit with high-set horizontal stabilizer for P1101.

14-17 Mar 1945 P1110 W armament studies for the P1112: Researcher: Büchler.

22-23 Mar 1945 Conference of the EHK in Bad Eilsen. No confirmed decision regarding the turbo-jet fighter in respect to the proposals of the TLR Chief, since Professor Messerschmitt as absent (Special Commission Day Fighter). TLR-F1.E (Knemeyer) granted the Junkers Firm a development contract for the EF-128 (though not a production series order).

24 Mar 1945   Design: high-set tail unit for the P1101. ("Experimentation Aircraft")

27 Mar 1945   Design XVIII/166 P1112-V1: conclusively "provisional" design.

End of Mar 1945   Conference for the turbo-jet fighter call for bids at Focke-Wulf in Bad Eilsen, by order of the RLM. The Tank Ta-183 would be declared the new standard fighter of the Luftwaffe, in memory of Professor Tank.

Beginning of Apr 1945 The P1101 is approximately 80% complete: though it still lacks a flight-ready engine.

8 Apr 1945   The British Army occupies the Focke-Wulf development department in Bad Eilsen.

18 Apr 1945   Outline for the P1112V1 mock-up.

29 Apr 1945   Invasion of the American infantry in Oberammergau.

Beginning of May 1945 Looting of the complex by the American infantry.

7 May 1945   Arrival of a delegation from the CAFT (Combined Advance Field Team): assessment of the significance of the research institute.

8 May 1945   Capitulation of the German Empire.

17 May 1945   The first group of Anglo-American experts arrive in Oberammergau.

21 May 1945   The co-founder and Technical Director of Bell Aircraft Works, R. J. Woods, arrives in the Upper Bavarian Research Institute, along with another group of specialists. With his co-workers, Hawkins and Haven, Woods leads the data collection and evaluation of the materials discovered.

June-July 1945   Interrogation of the Messerschmitt employees in the Haus "Osterbichl" hotel.

5 July 1945   A large portion of the documents hidden at the end of the war show up once again in Oberammergau.

Summer 1945   Attempt to recondition and completely build the P1101-V1 in Hangar 615: following this, the aircraft was disassembled and shipped to the USA.

Middle-End of May 1945 Immediately following evaluation of the captured German research documents, work begins on swept and tiltable wings in the NACA Research Center in Langley/Hampton for further development of the Bell X-1: with swept wing (= 40 degrees) - with forward-sweeping wing - with tiltable wing - with V-shaped tail unit John Campbell examines the tiltable wing in the free-flow wind tunnel at the Langley Research Center (LARC): the wings on the model could rotate from 0 to 60 degrees; surprisingly good flight characteristics with the sweep set at 40 degrees.

Oct 1945   Proposal: in place of the swept wing variant of the Bell X-1, construction of a new aircraft; the Bell X-2.

14 Dec 1945   Bell receives a development contract for two swept wing research aircraft for the Bell X-2.

20 July 1946   First swept wing aircraft in the United States: two rebuilt Bell P-63 A's inherit the designation L-39-1 (Number 90060) and L-39-2 (Number 90061); both aircraft possess a 35 degree wing sweep. Program is run under the leadership of the Navy; the wing data leads to the Douglas 558-2.

1 Aug 1946   Final report of the "research" into the Messerschmitt AG: Editor: Robert J. Woods of Air Technical Intelligence (ATI), Review Number F-IR-6 RE. Editor: Robert J. Woods Wright Field, Dayton, Ohio. Headquarters, Air Materiel Command.

1947   The research program "variable wing sweep" on a reworked Bell X-1 model at Langley. Charles Dunlan, Program Director proposes the future Bell X-2 tiltable wing.

3 July 1947	Official construction contract for the Bell X-2.
July 1948	Bell's construction team, under the direction of Robert J. Woods, proposes an improved, enlarged derivative of the P1101 with tiltable wing. Beginning of the X-5 program. For the U.S. Air Force, the design was proposed as an interceptor. However, later, unfavorable reports stopped the planned construction of 24 aircraft.
Aug 1948	Inclusion of the infinished P1101 prototypes into the X-5 program. During transport of the German aircraft from Wright Air Force Base in Dayton, Ohio, to Bell Aircraft Works in Buffalo, New York, the P1101 fell off the ransport vehicle and was — once again — heavily damaged.
Feb 1949	Beginning of the wind tunnel tests for the X-5 program in Langley.
9 May 1949	The Power Plant Laboratory, Wright Field, asserts reservations on the configuration of the fuel being directly over the engine.
13 July 1949	Signing of the contract for construction of two X-5 test aircraft: contract number W-33-038-ac-3298.
29 Aug 1949	Wright Field Aircraft Laboratory proposes several extensive modifications on the basic design of the X-5 and questions the sense behind the entire program.
Late Autumn 1949	Completion of the Bell X-5 wood mock-up.
Beginning of Dec 1949	From the inspection of the mock-up, there are 76 desired modifications and proposals: 40 of these concern Bell Aircraft Works.
Beginning of 1951	Roll-out and inclusive ground testing of the first aircraft, 50-1838.
March 1951	Messerschmitt makes contact with the Spanish Hispano Aviacion (HA) aircraft company in Seville, Spain.
9 June 1951	Transfer of the X-5/50-1838 inside a C-119 from Bell Works to Edwards Air Force Base, California.
20 June 1951	Maiden flight of the Bell X-5/50-1838. Pilot: Jean Ziegler; world premiere of the wing with variable sweep.
27 June 1951	During the fifth flight, manipulation of the tiltable wing takes place.
17 July 1951	Second visit for Messerschmitt in Spain: memorandum to the Spanish Ministry of Aviation.
8 Oct 1951	Final flight of the first test aircraft by Bell Works with impending transfer to the U.S. Air Force.
26 Oct 1951	First consultation contract between Professor Messerschmitt and HA.
7 Nov 1951	Formal transfer of the 50-1838 to the U.S. Air Force. NACA begins the evaluation program.
10 Dec 1951	Maiden flight of the second prototype Bell X-5/50-839: Pilot: once again, Jean Ziegler. During the ninth flight, the wing is fully swept (from a 20 degree sweep to 60 degrees and back again).
18 Dec 1951	The second test aircraft, 50-1839, is accepted by the U.S. Air Force.
3 Jan 1952	Work begins for the first Messerschmitt Development Team in Seville.
27 June 1952	After a series of difficulties, the first glide flight for the Bell X-2. First flight of the Bell X-2 with rocket engine in operation on November 18, 1955.

13 Oct 1953	Crash of the second Bell X-5 prototype during a spin test: Pilot Raymond Popson died during this crash. With the remaining test aircraft, 50-1838, the evaluation program will be continued until 1955: the variable-sweep wing showed the expected characteristics during evaluations.

Notes

21 Dec 1964	Maiden flight of General Dynamic's F-111A: introduction to a new era in aircraft construction: breakthrough of the variable-wing sweep aircraft.
1967	Maiden flight of the Soviet Sukhoi Su-17 with variable-sweep wings according to the patent from Dr. Alexander Lippisch.
21 Dec 1979	Maiden flight of the slant wing test aircraft, NASA/Ames AD-1. This aircraft is the realization of the unseparated variable geometric wing from Dr. Richard Vogt.
	Thereby, the ideas and theoretical work of the German forward thinkers (Messerschmitt/ Voigt, Lippisch and Vogt) was realized to the full extent and also successfully tested.

Beginning of 1954	Messerschmitt's Development Bureau in Seville constructs the first designs for Project 300. Wind tunnel testing begins in Switzerland.
2nd half of 1954	Increased work on the P300 wind tunnel evaluations in Emmen, Switzerland and at the INTA in Torrejon, Spain, near Madrid.
24 May 1955	Modifications to the P-300's present concept: Determination for the second phase of development.
13 July 1955	Messerschmitt's first considerations to a smaller P300: actual start of the development of the HA-300 light fighter.

23 Nov 1955	Specification for the HA-300 light fighter.
7 Dec 1955	Messerschmitt's design for a "micro fighter."
8 Feb 1956	Memorandum on the P-300 supersonic fighter design.
7 Mar 1956	Project Hand over I. Delta wing with a relative thickness ratio of 3%.
End of 1956	Construction of the engineless glider completed.
1957-1958	Financial problems for Spain hinder construction of the factory designs. Construction work is carried out with sharply reduced team size.
25 Jun 1959	First and only take-off of the HA-300 P glider from the San Pablo airfield in two of a Spanish He 111.
End of 1959	Spain sells several HA-300 documents to the United Arab Republic (Egypt).
Beginning of 1960	Bristol halts development of the Br. Or. 12 power plant: Egypt enlists F. Brandner to develop an engine for the HA-300.
May 1960	Construction of a horizontal stabilizer for the HA-300.
Sep 1960	Subsonic testing in the Swiss Aircraft Works wind tunnel in Emmen.
Summer 1961	Supersonic tests in the 8 foot (2.44m) wind tunnel in Bedford in Bedfordshire, England.
1961-1962	Further development until construction-ready status; construction of factory designs; assembly of full design.
31 Oct 1962	Specifications for the HA-300 from the Messerschmitt Study Bureau.

July 1963	First bench running of the HA-300's E-300 engine.
23 July 1963	The HA 300.001 prototype taxis past Egyptian President Nasser during Egyptian National Holiday parade.
Feb 1964	Resumption of the taxiing tests following a few modifications to the 001.
7 Mar 1964	Maiden flight of the HA-300. Egypt's financial and domestic politic problems hinder the program and turn back the production schedule.
1964-1965	Flight test with the 001 and breaking-in of the modifications.
22 July 1965	Maiden flight of the HA-300.002.
June-July 1966	Initiation of flight testing with the E-300 power plant in an Antonov An-12 and an Indian HF-24 "Maruf."
June 1967	Six-Day War with Israel.
Summer 1967	End of Messerschmitt's involvement in Egypt: Professor Tank, arriving from India, takes his place.
May 1969	Egypt discontinues the program and the foreign workers are forced to leave the country by June 1, 1969.
Nov 1969	Taxiing tests with the HA-300.003, not equipped with the fully operational E-300. There were no more flight tests.

Authors' Notes

The documents used originated mainly from the archives of the authors.

Our special thanks to the former employees of Messerschmitt AG, without whose energetic cooperation the undertaking would hardly have been possible. The representatives named here are Hans Kaiser and Wolfgang Degel.

Further, our thanks to:

Manfred Böhme, Hanfried Schliephake, Jay Miller, Karl Pawlas, Manfred Griehl, David Myhra, Hans Redemann, and above all, Günther Sengfelder, whose outstanding construction model helped illustrate this book.

We further thank the Deutsches Museum in Munich and the Musée de l'Air et de l'Espace in Paris, Le Bourget.